T0145322

Engineering Materials

This series provides topical information on innovative, structural and functional materials and composites with applications in optical, electrical, mechanical, civil, aeronautical, medical, bio- and nano-engineering. The individual volumes are complete, comprehensive monographs covering the structure, properties, manufacturing process and applications of these materials. This multidisciplinary series is devoted to professionals, students and all those interested in the latest developments in the Materials Science field.

More information about this series at http://www.springer.com/series/4288

T. Daniel Thangadurai · N. Manjubaashini ·
Sabu Thomas · Hanna J. Maria

Nanostructured Materials

T. Daniel Thangadurai
Department of Nanoscience
and Technology
Sri Ramakrishna Engineering College
Coimbatore, Tamil Nadu, India

N. Manjubaashini
Department of Nanoscience
and Technology
Sri Ramakrishna Engineering College
Coimbatore, Tamil Nadu, India

Sabu Thomas
IIUCNN
Mahatma Gandhi University
Kottayam, Kerala, India

Hanna J. Maria
IIUCNN
Mahatma Gandhi University
Kottayam, Kerala, India

ISSN 1612-1317 ISSN 1868-1212 (electronic)
Engineering Materials
ISBN 978-3-030-26147-4 ISBN 978-3-030-26145-0 (eBook)
https://doi.org/10.1007/978-3-030-26145-0

This Springer imprint is published by the registered company Springer Nature Switzerland AG
The registered company address is: Gewerbestrasse 11, 6330 Cham, Switzerland

Contents

Chapter 1
Nanotechnology and Dimensions

Abstract The role of nanotechnology towards different field of application emphases by nanomaterials size, dimensions and properties. Varying nanometer size and dimensions leads to different nanostructures with unique properties, and the features of nanoparticles was identified by their synthesis route and analyses techniques. Also the chapter briefly discussed about significances and concepts of nanoscience and technology.

1.1 Fundamentals of Nanomaterials

The design, construction and utilization of functional structures with at least one characteristic dimension measured in nanometers are designated as nanomaterials. Thus, the nanomaterials emphasize in the field of science, engineering and technology, which produces the way for nanotechnology founding. These nanomaterials show dramatic change in their physical, chemical and biological properties due to its reduced size and shape [1].

Nanoscience and nanotechnology predominantly deal with the synthesis, characterization, investigation, and use of nanostructured materials. These materials are characterized by at least one dimension in the nanometer (1–100 nm) range. Nanostructures create a link between molecules and bulk materials. The nanostructures include clusters, quantum dots, nanocrystals, nanowires, nanotubes, arrays, assemblies, and superlattices of the specific nanostructures. The physical and chemical properties of nanomaterials can change significantly from those of the atomic-molecular or the bulk materials of the same structure. The uniqueness of the structural characteristics, energetics, response, dynamics, and chemistry of nanostructures creates the basis of nanoscience. Appropriate control of the properties and reaction of nanostructures can lead to new devices and technologies [2]. Thus, the electron states of nanostructures are leading to new electrical, thermal, magnetic, optical, and mechanical properties at the nanoscale. Nanostructures physico-chemical properties are tuned by governing their size and shape at the nanoscale, which leads to unique applications. But, the physico-chemical properties of the nanostructures are limited only by the precision of the experimental techniques used for the fabrication

T. D. Thangadurai et al., *Nanostructured Materials*, Engineering Materials,
https://doi.org/10.1007/978-3-030-26145-0_1

of the low-dimensional structures. The energy spectrum of a quantum well, quantum wire, or quantum dot can be engineered by controlling (i) the size and shape of the confinement region and (ii) the strength of the confinement potential [3].

1.2 Dimensions of Nanomaterials

Nanotechnology deals with materials in zero, one, two and three dimensions as nanometers and the nanostructured matcrials. The physical and chemical properties of materials undergo exciting changes in their sizes of nanometer dimension. Most biomolecules and bioentities are of nanometer size; thus the nanoscale provides the great prospect to study such bioentities and their interactions with other materials [4]. Another invitation is semiconductor industry for its ever-lasting demand for miniaturization, has been compelled deeply into the nano-realm.

The concept of quantum confinement arises from the size reduction of material leading to the electronic wave functions being more tightly confined and resulting in changes associated with electronic and optical properties of the nanomaterial. A smaller (or bigger) particle size results in a stronger (or weaker) confinement which gives rise to enhancement (or decrease) of the band gap and modifies the band structure of the material. This results in changes in the electron mobility and effective mass, relative dielectric constant, optical properties to name a few. For the reason of quantum confinement effect metallic nanoparticles show interesting properties like variations in color of colloidal suspensions with changing particle size, UV photoemission, enhanced photoluminescence, etc. The light-induced UV emission in Ag nanoparticles has prompted the use of nano-Ag coatings for antimicrobial purposes in the healthcare sector [5].

1.2.1 2D Confinement

Thin films with thickness of few nanometers are usually deposited on a bulk material. Their properties may be dominated by surface and interface effects or they reflect the confinement of electrons in the direction perpendicular to the film. In the two dimensional confinement which is parallel to the film the electrons behave like in a bulk material [6].

1.2.2 1D Confinement

Nanolaminated or compositionally modified materials, Grain boundary films, Clay platelets, Semiconductor quantum wells and superlattices, Magnetic multilayers and spin valve structures, Langmuir–Blodgett films, Silicon inversion layers in field

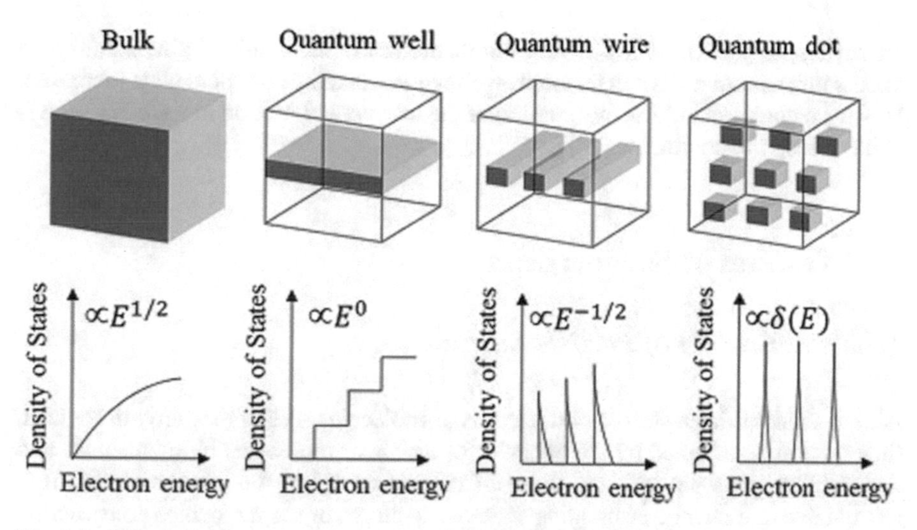

Fig. 1.1 Schematic illustration of quantum effects from bulk, quantum well, quantum wire to QD. Reprinted with permission from © Journal of Semiconductors and Semimetals, 101, 305–354 (2019)

effect transistors, Surface-engineered materials for increased wear resistance or corrosion resistance [7]. The size of nanoparticles becomes smaller than the de Broglie wavelength, the mean free path of electrons acts as quantum dots. The band gap increases, and a continuous density of states (DOS) appear in the case of bulk materials (Fig. 1.1a) there are discrete energy levels, similar to the case of atoms. This has a considerable influence the optical properties of nanoparticles, as the absorption shifts from the infrared to the visible range. Quantum confinement in quantum wells (one-dimensional), quantum wires (two-dimensional) and quantum dots (three-dimensional) and the resulting changes of the energy spectrum as shown in figure has serious implications for the optoelectronic properties of nanostructured semiconductors, and is currently utilized in optoelectronic devices [8].

1.2.3 Zero Dimensional Confinement

In a zero-dimensional structure, the exciton states are determined by a competition between the confinement effects and the correlation of the electron-hole pair induced by the coulomb interaction. The single-particle levels, with splittings which scale as $1/L^2$ where L is the size of the quantum box, which are determined by the confinement. The coulomb energies scale as l/L. As the confinement in a zero-dimensional structure increases, the coulomb-induced mixing of states becomes more difficult, even though the Coulomb energies increase, and electrons and holes become frozen in the lowest energy single-particle states. Freeze-out of motion in the confined dimensions also occurs in quantum wells and quantum-well wires when the confinement is increased. However, coulomb induced correlation of the unconfined motion in a well or wire is enhanced by the confinement [9].

1.3 Features of Nanoparticles

1.3.1 Activation of Particle Surface

All the solid particles consist of the atoms or molecules. When they are micronized, they tend to be affected by the behavior of atoms or molecules by themselves and show different properties from those of the bulk solid of the same material. It is attributable to a change of bonding state of the atoms or the molecules constructing the particles. For example, if a cube with a side length of 1 cm is divided into a cube of 1 μm, the particle number increases to 10^{12} and being divided into the one of 10 nm, then it amounts to 10^{18}, where the fraction of the atoms or the molecules located at the surface on the particles plays a great role, since they are more active than those inside the solid particles because of the free hand, which leads to easy bonding with the contacting materials and causes various changes in particle properties. The diameter of the smallest hydrogen atom is 0.074 nm, and that of the relatively large lead atom is 0.35 nm. From these sizes, it is estimated that the particle with a size of 2 nm consists of only several tens to thousands atoms. When the particle is constructed by larger molecules, the number decreases furthermore. On the other hand, as the micronization of solid particles, the specific surface area increases generally in reversal proportion to the particle size. In the above-mentioned case, when the particle of 1 cm is micronized to 1 μm and 10 nm, the specific surface area becomes ten thousand times and million times, respectively. As the increase in the specific surface area directly influences such properties like the solution and reaction rates of the particles, it is one of major reasons for the unique properties of the nanoparticles different from the bulk material together with the change in the surface properties of the particles itself.

1.3.2 Particle Size

Particle size is tenure to represent the three-dimensional particle in one-dimensional scalar value. The size of any spherical particle can be signified by its diameter with no obscurity. The geometric size is obtained through three dimensional measurements of a particle to get its width, thickness and length, and then calculating one-dimensional value such as arithmetic mean. In practice, one-dimensional value obtained based on the two dimensional-projected outlines is utilized such as diameter of a circle having the same area as the projected area. Statistical diameter based on one-dimensional measurement is also well applied in practice such as a Feret diameter, which is determined as the distance between pairs of parallel tangents to the particle silhouette in some fixed direction. As for the equivalent size in relation to practical methods of particle size measurements, there are many different definitions such as sieve diameter based on sieving, equivalent light-scattering diameter, Stokes diameter based on particle motion in fluid, and the equivalent diameter based on the Brownian motion. These equivalent diameters give, usually, different values depending on the measurement principles unless the particles are spherical. Specific surface area of powder or equivalent-specific surface diameter is well applied to the evaluation of nanopowders. However, in this case, the particles should not be porous. Besides, the method cannot be applied to get particle size distribution.

1.3.3 Particle Shape

The essential properties such as particle diameter, particle shape of nano-size or fine particles influence the character of the particle-packed bed. In these properties, the particle diameter measurement equipment based on various principles can be promoted, and it is easy to measure particle diameter distribution. But a particle shape analyzer for nanoparticle cannot be easily found and the shape index of nanoparticles can be calculated from particle images observed using various types of microscopes.

1.3.4 Two-Dimensional Particle Projection Image

In order to measure the particle shape, the outline of particle shape of two-dimensional projection images captured from microscopic photograph is analyzed. Since the diameter of a nanoparticle is smaller than the wavelength of visible light, a nano-size particle cannot be observed by an optical microscope. Usually, the two-dimensional projection image of nanoparticles is captured by scanning electron microscope (SEM) or transmission electron microscope (TEM), and the particle shape indices can be calculated from the captured images by image analysis software. For the shape analysis of fine particle over micrometer order, automatic particle shape analyzers using two-dimensional image of particles in a sheath flow are available. This analyzer captures particle images automatically by an optical microscope with CCD camera under stroboscope flush lighting.

1.3.5 Three-Dimensional Particle Image

In the particle shape measurement of flaky particle or porous particle including hole or space inside the particle, the shape analysis of a two-dimensional particle projection picture is inadequate, and three-dimensional shape analysis is necessary. Although it is difficult to measure the thickness of a particle by the ordinary electron microscope, the thickness or surface roughness can be measured by three-dimensional scanning electron microscope (3D-SEM). 3D-SEM takes two microscopic pictures from slightly different angles and obtains the three dimensional information including thickness and surface roughness geometrically. However, if one of the pictures has the hidden area in the shadow of a particle, the three-dimensional information of the area is hidden. When well dispersed nanoparticles adhere on a flat substrate, the height difference between the particle surface and the flat substrate can be measured accurately, and the thickness of nanoparticle is obtained. Using the TEM, 120 transmission images are taken when a sample is rotated 1° interval from −60° to +60°. This three-dimensional imaging technique called TEM-CT, which is similar to the computer aided tomography, is expected to be applied for detailed three-dimensional shape measurement of nanoparticles. Using the scanning probe microscope (SPM) including atomic force microscope (AFM), the surface roughness can be measured by tiny probe with the high resolution under nano-meter order. SPM are very effective for thickness measurement of nanoparticles. The vertical direction length such as particle thickness can be measured accurately by SPM, but the measured horizontal length becomes bigger about the diameter of the probe. It means that the particle diameter in horizontal direction measured by SPM is bigger than real particle size. Moreover, for a soft particle or an adhesive particle, a particle position changes by contact of probe or a particle adheres to probe, and an accurate image is not obtained [10]. The major interest in nanoparticles has been generated in the research and the industry sector mostly due to its physical and chemical properties which differ quite a lot from the bulk (large scale) materials. At present almost all branches of applied science and technology are being advanced in some way or the other with the advent of nanotechnology.

The surface to volume ratio (S/V) is simply the surface area of an object divided by its volume. For symmetrical objects like a spherical particle this ratio is inversely proportional to the radius, and for a cube it is inversely proportional to its sides. The increasing S/V more and more atoms/molecules of the material become exposed to the surroundings and a larger number of the so-called "dangling bonds" become available at the surface, thus making the particle more active chemically. For a very simple example a spherical particle of ~100 nm diameter has only a very small percentage ~2% of its total constituent atoms exposed to the surface, if we go down to particle diameters of ~10 nm 20–25% of the atoms become exposed to the surroundings, while a further size reduction say down to 3 nm increases the fraction of atoms exposed at the particle surface to a 45–60%. In such conditions the behavior of the particle becomes altered in terms of its chemical activity. This is precisely the reason behind certain noble metals like Ag, Au, Pt, etc., behaving as highly potent catalysts in their particulate, especially nanoparticulate form. The presence of large number of unsaturated dangling bonds on the graphene (a two-dimensional sp^3 bonded honeycomb-like sheet of carbon atoms, which is the building block of graphite) surface which has a very high S/V makes it an excellent sensing material for chemical and gas detection. Such high S/V ratio is also responsible for the lowering of melting point in nanocrystals [11].

1.4 Significances of Nanotechnology

The nanotechnology exactly means any expertise performed on a nanoscale that has applications in the real world. The nanotechnology incorporates the production and application of physical, chemical, and biological systems at scales ranging from individual atoms or molecules to submicron dimensions as well as the combination of the resulting nanostructures into larger systems. The nanotechnology is possible to have an intense impression on our economy and society in the early twenty first century, similar to that of semiconductor technology, information technology or cellular and molecular biology. Science and technology research in nanotechnology potentials advances in areas as materials and manufacturing, medicine and healthcare, energy, biotechnology, information technology and national security. It is widely felt that nanotechnology will be the next industrial revolution [12].

Nanometer scale structures are primarily built up from their elemental constituents. Chemical synthesis—the spontaneous self-assembly of molecular clusters from simple reagents in solution or biological molecules are used as building blocks for the production of three-dimensional nanostructures, including quantum dot of arbitrary diameter. A variety of vacuum deposition and nonequilibrium plasma chemistry techniques are used to produce layered nanocomposites and nanotubes. Atomically controlled structures are produced using molecular beam epitaxy and organo-metallic vapor phase epitaxy. Micro and nanosystem components are fabricated using top-down lithographic and non-lithographic fabrication techniques and range in size from micro to nanometers. Continued improvements in lithography for use in the production of nano components have resulted in line widths as small as 10 nm in experimental prototypes. The nanotechnology field, in addition to the fabrication of nanosystems provides the impetus to development of experimental and computational tools.

The micro and nanosystems contain micro/nano electromechanical systems, micro mechatronics, optoelectronics, and microfluidics and systems integration. These systems can sense, control and activate on the micro/nanoscale and function individually or in arrays to generate effects on the macroscale. Due to the permitting nature of these systems and major influence they can have on the commercial and defense applications, venture capitalists, industry, as well as the federal government have taken a distinct attention in nurturing growth in this field. Micro and nanosystems are likely to be the next logical step in the silicon revolution. Science and technology continue to move forward in making the fabrication of micro/nano devices and systems possible for a variety of industrial, consumer, and biomedical applications. A range of MEMS devices have been produced, some of which are commercially used. A variety of sensors are used in industrial, consumer, and biomedical applications. Various microstructures or micro components are used in micro instruments and other industrial applications such as micro mirror arrays. Integrated capacitive type, silicon accelerometers have been used in airbag deployment in automobiles. Other major industrial applications include pressure sensors, inkjet printer heads, and optical switches. Silicon-based piezoresistive pressure sensors for manifold absolute pressure sensing for engines were launched in 1991 by Nova-Sensor and their annual sales were about 25 million units in 2002. Annual sales of inkjet printer heads with microscale functional components were about 400 million units in 2002. Capacitive pressure sensors for tire pressure measurements were launched by Motorola. Other applications of MEMS devices include chemical sensors, gas sensors, infrared detectors, and focal plane arrays for earth observations, space science and missile defense applications, pico-satellites for space applications, and many hydraulic, pneumatic, and other consumer products. MEMS devices are also being pursued in magnetic storage systems, where they are being developed for super-compact and ultrahigh recording-density magnetic disk drives. Several integrated head/suspension micro devices have been fabricated for contact recording applications. High bandwidth, servo-controlled micro actuators have been fabricated for ultrahigh track density applications which serve as the fine position control element of a two-stage, coarse/fine servo system coupled with a conventional actuator. Millimeter-sized wobble motors and actuators for tip based recording schemes have also been fabricated [13].

1.5 Basic Concept of Nanotechnology

The nanotechnology is a study of structure between 1 and 100 nm in size. The nanotechnology is concerned with materials and systems whose structure component exhibit novel and significantly improved physical, chemical, and biological properties, phenomena and process because of their nanoscale size. The structural features in the range of ~10^{-9} to 10^{-7} m i.e. 1–100 nm determine important changes as compared to the behavior of isolated molecules or of bulk materials. It is an interdisciplinary science involving physics, chemistry, biology, engineering materials science, computer science etc. [14].

Nanostructures are described as novel materials whose size of elemental structure has been plotted as nanometer scale. Nanomaterials may be classified on the basis of dimensionality and modulation. Some special nanostructures like nanotubes, nanoporous materials, aerogels, zeolites, core-shell structures have also come up with their novel characteristics. A number of methods have been used for the synthesis of nanostructure with various degrees of success and many direct and indirect techniques are employed for their characterization. The fact, which makes the nanostructures interesting, is that the properties become size dependent in nanometer range because of surface effect and quantum confinement effect. The geometric structure, chemical bonds, ionization potential, electronic properties, optical properties, mechanical strength, thermal properties, magnetic properties etc. are all affected by particle sizes in nanometer range [15]. Nanomaterials exhibit properties often superior to those of conventional coarse-grained materials. These include increased strength, enhanced diffusivity, improved ductility, reduced density, reduced elastic modulus, higher electrical resistivity, increased specific heat, higher thermal expansion coefficient, lower thermal conductivity, increased oscillator strength, blue shifted absorption, increased luminescence and superior soft magnetic properties in comparison to conventional bulk materials. Use of nanostructured materials has produced transistors with record low speed and lasers with low threshold current. These are being used in compact disk player systems, low noise amplification in satellite receivers as sources for fiber optic communication etc. Beneficial applications of nanomaterials include self-cleaning glass, UV resistant wood coating etc. Nanoscale devices are being used in medical field also for diagnosis, treatment and prevention of diseases and in drug delivery system, magnetic resonance imaging, radioactive tracers etc. [16].

References

1. Kelsall RW, Hamley IW, Geoghegan M (2005) Nanoscale science and technology. Wiley, England
2. Ali Mansoori G, Fauzi Soelaiman TA (2005) Nanotechnology—an introduction for the standards community. JAI 2:1–21
3. Kurzydlowski KJ (2006) Physical, chemical, and mechanical properties of nanostructured materials. Mater Sci 42:85–94
4. Kaur G, Singh T, Kumar A (2012) Nanotechnology: a review. IJEAR 2:50–53
5. Alivisatos AP (1996) Semiconductor clusters, nanocrystals, and quantum dots. Science 271:933–937
6. Mu P, Zhou G, Chen CL (2018) 2D nanomaterials assembled from sequence-defined molecules. Nano-Struct Nano-Objects 15:153–166
7. Poole CP, Owens FJ (2003) Introduction to nanotechnology. Wiley, Hoboken, NJ
8. Altavilla C, Ciliberto E (2011) Inorganic nanoparticles: synthesis, applications, and perspectives. Taylor and Francis Group, LLC, Boca Raton, FL
9. Bryant GW (1990) Understanding quantum confinement in zero-dimensional nanostructures: optical and transport properties. In: Beaumont SP, Torres CMS (eds) Science and engineering of one- and zero-dimensional semiconductors. NATO ASI series (series B: physics), vol 214. Springer, Boston, MA
10. Hosokawa M, Nogi K, Naito M et al (2007) Nanoparticle technology handbook. Elsevier, Netherlands
11. Sengupta A, Sarkar CK (2015) Introduction to nano basics to nanoscience and nanotechnology. Springer, Heidelberg, New York
12. Stirling DA (2018) The nanotechnology revolution. A global bibliographic perspective. Pan Stanford Publishing, Singapore
13. Purohit K, Khitoliya P, Purohit R (2012) Recent advances in nanotechnology. IJSER 3:1–10
14. Behari J (2010) Principle of nanoscience: an overview. Indian J Exp Biol 48:1008–1019
15. Zhang X, Cheng X, Zhang Q (2016) Nanostructured energy materials for electrochemical energy conversion and storage: a review. J Energy Chem 3:1–18
16. Mandal G, Ganguly T (2011) Applications of nanomaterials in the different fields of photosciences. Indian J Phys 85:1229–1245

Chapter 2
Nanomaterials, Properties and Applications

Abstract Nanomaterials size range from 1 to 100 nm with unique optical, electrical, magnetic, mechanic and structural properties. There are some naturally occurring nanomaterials (viruses, protein, lotus leaf, spider-mite silk, and butterfly wings etc.) and more engineered nanomaterials (Au NPs, Ag NPs, etc.) with top-down and bottom-up approaches. The chapter detailed about properties and applications of nanomaterials.

2.1 Brief Notes on Nanomaterials

Nanomaterials are keystones of nanoscience and nanotechnology. Nanostructure science and technology is a wide and interdisciplinary area of research and development activity that has been rising explosively worldwide in the past few years. It has the potential for revolutionizing the ways in which materials and products are created and the range and nature of functionalities that can be accessed [1]. It already has important viable influence, which will certainly increase in the future. Nanoscale materials are defined as a set of substances where at least one dimension is less than approximately 100 nm. A nanometer is one millionth of a millimeter—approximately 100,000 times smaller than the diameter of a human hair. Nanomaterials are of interest because at this scale unique optical, magnetic, electrical, and other properties emerge. These emergent properties have the potential for great impacts in electronics, medicine, and other fields [2].

Some nanomaterials occur naturally, but of particular interest are Engineered Nanomaterials (EN), which are designed and being used in many commercial products and processes. They can be found in sunscreens, cosmetics, sporting goods, stain resistant clothing, tires, electronics, and used in medicine for purposes of diagnosis, imaging and drug delivery. The nanoparticles are actually a fairly common type of material in many different environments, and they can pass by almost undetected unless you are looking for them. The physical and chemical processes produce nanoparticles. Naturally occurring nanoparticles can be found in volcanic ash, ocean spray, fine sand and dust, and even biological matter [3] (Fig. 2.1).

T. D. Thangadurai et al., *Nanostructured Materials*, Engineering Materials,
https://doi.org/10.1007/978-3-030-26145-0_2

Fig. 2.1 Naturally occurring and man-made nanoparticles

Synthetic nanoparticles are derived from two general categories, incidental and engineered nanoparticles. Incidental nanoparticles are the byproducts of human activities, generally have poorly controlled sizes and shapes, and may be made of a hodgepodge of different elements. Many of the processes that generate incidental nanoparticles are common every day activities: running diesel engines, large-scale mining, and even starting a fire. Engineered nanoparticles on the other hand, have been specifically designed and deliberately synthesized by human beings. Not surprisingly, they have very precisely controlled sizes, shapes, and compositions. They may even contain layers with different chemical compositions. Although engineered nanoparticles get more sophisticated with each passing year, simple engineered nanoparticles can be created by relatively simple chemical reactions that have been within the scope of chemists and alchemists for many centuries [4] (Fig. 2.2).

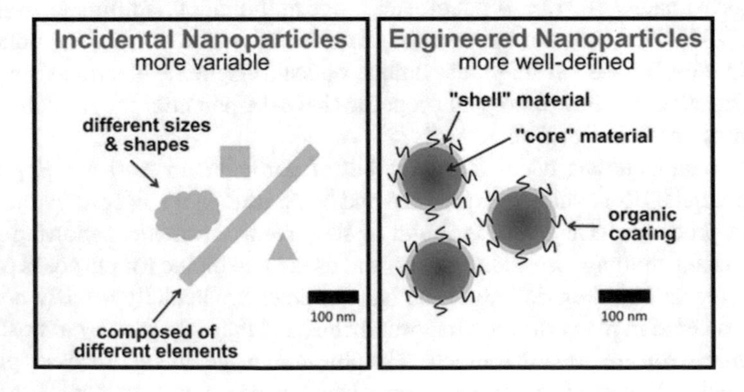

Fig. 2.2 Incidental and engineered nanoparticle

2.2 Nanomaterial Properties

2.2.1 *Structural Properties*

The increase in surface area and surface free energy with decreasing particle size leads to changes in interatomic spacing. For Cu metallic clusters the interatomic spacing is detected to decrease with decreasing cluster size. This effect can be described by the compressive strain induced by the internal pressure arising from the small radius of curvature in the nanoparticle. Equally, for semiconductors and metal oxides there is indication that interatomic spacings increase with decreasing particle size. The apparent stability of metastable structures in small nanoparticles and clusters, such that all traces of the usual bulk atomic arrangement become lost. Metallic nanoparticles, such as gold, are known to adopt polyhedral shapes such as cubeoctahedra, multiply twinned icosahedra and multiply twinned decahedra. These nanoparticles may be regarded as multiply twinned crystalline particles in which the shapes can be understood in terms of the surface energies of various crystallographic planes, the growth rates along various crystallographic directions and the energy required for the formation of defects such as twin boundaries. However, there is convincing indication that such particles are not crystals but are quasiperiodic crystals or crystalloids. These icosahedral and decahedral quasicrystals form the basis for further growth of the nanocluster, up until a size where they will switch into more regular crystalline packing arrangements [5].

Crystalline solids are different from amorphous solids in that they possess long range periodic order and the patterns and symmetries which occur correspond to those of the 230 space groups. Quasiperiodic crystals do not possess such long-range periodic order and are distinct in that they exhibit fivefold symmetry, which is forbidden in the 230 space groups. In the cubic close-packed and hexagonal close-packed structures, exhibited by many metals, each atom is coordinated by 12 neighbouring atoms. All of the coordinating atoms are in contact, although not evenly distributed around the central atom. However, there is an alternative arrangement in which each coordinating atom is situated at the apex of an icosahedron and in contact only with the central atom. The rigid atomic sphere model allow the central atom to reduce in diameter by 10%, the coordinating atoms come into contact and the body now has the shape and symmetry of a regular icosahedron with point group symmetry 235, indicating the presence of 30 twofold, 20 threefold and 12 fivefold axes of symmetry.

This geometry represents the nucleus of a quasiperiodic crystal which may grow in the forms of icosahedra or pentagonal dodecahedra. These are dual solids with identical symmetry, the apices of one being replaced by the faces of the other. Such quasiperiodic crystals are known to exist in an increasing number of aluminium-based alloys and may be stable up to microcrystalline sizes. It should be noted that their symmetry is precisely the same as that of the fullerenes C_{20} and C_{60}. Hence, as fullerenes, quasiperiodic crystals are expected to have an important role to play in nanostructures. The size-related instability characteristics of quasiperiodic crystals

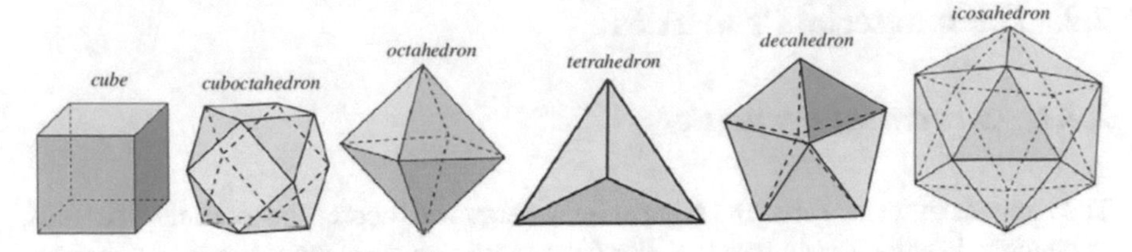

Fig. 2.3 Geometrical shapes of cubo-octahedral particles and multiply twinned decahedral and icosahedral particles. Reprinted with permission from © Advances in Natural Sciences: Nanoscience and Nanotechnology, 1 (2010)

are not well assumed. An often detected process appears to be that of multiple twinning, such crystals being distinguished from quasiperiodic crystals by their electron diffraction patterns. Here the five triangular faces of the fivefold symmetric icosahedron can be mimicked by five twin-related tetrahedral through relatively small atomic movements [6] (Fig. 2.3).

2.2.2 *Thermal Properties*

The large increase in surface energy and the change in interatomic spacing as a function of nanoparticle size have an effect on material properties. For instance, the melting point of gold particles, which is really a bulk thermodynamic characteristic, has been observed to decrease rapidly for particle sizes less than 10 nm, as shown in Fig. 2.4. There is evidence that for metallic nanocrystals embedded in a continuous matrix the opposite behavior is true; i.e., smaller particles have higher melting points [7].

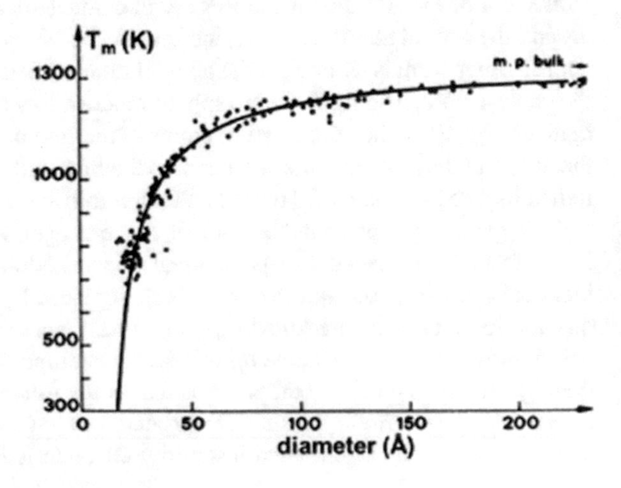

Fig. 2.4 Schematic diagram of the variation in melting point of gold nanoparticles as a function of particle size. Reprinted with permission from © Handbook of Nanoparticles, Springer, Cham (2016)

2.2.3 Chemical Properties

The change in structure as a function of particle size is intrinsically linked to the changes in electronic properties. The ionization potential is generally higher for small atomic clusters than for the corresponding bulk material. Furthermore, the ionization potential exhibits marked fluctuations as a function of cluster size. Such effects appear to be linked to chemical reactivity, such as the reaction of Fen clusters with hydrogen gas (Fig. 2.5). Nanoscale structures such as nanoparticles and nanolayers have very high surface area to volume ratios and potentially different crystallographic structures which may lead to a radical alteration in chemical reactivity. Catalysis using finely divided nanoscale systems can increase the rate, selectivity and efficiency of chemical reactions such as combustion or synthesis whilst simultaneously significantly reducing waste and pollution. Gold nanoparticles smaller than about 5 nm in diameter are known to adopt icosahedral structures rather than the normal face centered cubic arrangement. This structural change is accompanied by an extraordinary increase in catalytic activity. Furthermore, nanoscale catalytic supports with controlled pore sizes can select the products and reactants of chemical reactions based on their physical size and thus ease of transport to and from internal reaction sites within the nanoporous structure. Additionally, nanoparticles often exhibit new chemistries as distinct from their larger particulate counterparts; for example, many new medicines are insoluble in water when in the form of micron-sized particles but will dissolve easily when in a nanostructured form [8].

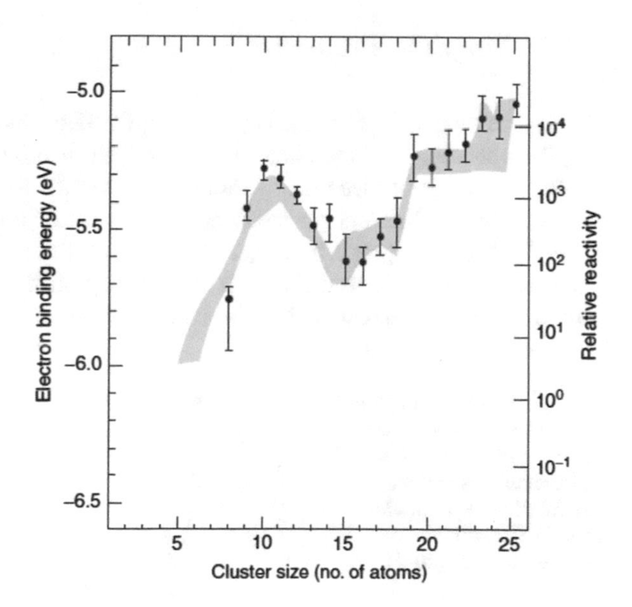

Fig. 2.5 Schematic diagram of the dependence of the electron binding energy and relative chemical reactivity of iron clusters to hydrogen gas as a function of cluster size. Reprinted with permission from © Condensed Matter & Biological Systems, 1, 1–36 (2010)

2.2.4 Mechanical Properties

The mechanical properties like toughness are highly dependent on the ease of formation or the presence of defects within a material. As the system size decreases, the ability to support such defects becomes increasingly more difficult and mechanical properties will be altered accordingly. Novel nanostructures, which are very different from bulk structures in terms of the atomic structural arrangement, will obviously show very different mechanical properties. For example, single and multi-walled carbon nanotubes show high mechanical strengths and high elastic limits that lead to considerable mechanical flexibility and reversible deformation. As the structural scale reduces to the nanometer range, for example, in nanolayered composites, different scale dependence from the usual Hall–Petch relationship for yield strength often becomes apparent with large increases in strength reported. In addition, the high interface to volume ratio of consolidated nanostructured materials appears to enhance interface-driven processes such as plasticity, ductility and strain to failure. Many nanostructured metals and ceramics are observed to be superplastic, in that they are able to undergo extensive deformation without necking or fracture. This is presumed to arise from grain boundary diffusion and sliding, which becomes increasingly significant in a fine-grained material. Overall these effects extend the current strength–ductility limit of conventional materials, where usually a gain in strength is offset by a corresponding loss in ductility [9].

2.2.5 Magnetic Properties

Magnetic nanoparticles are used in many applications, like ferro-fluids, colour imaging, bioprocessing, refrigeration as well as high storage density magnetic memory media. The large surface area to volume ratio results in a substantial proportion of atoms having a different magnetic coupling with neighbouring atoms, leading to differing magnetic properties. Figure 2.6 shows the magnetic moments of nickel nanoparticles as a function of cluster size. Whilst bulk ferromagnetic materials usually form multiple magnetic domains, small magnetic nanoparticles often consist of

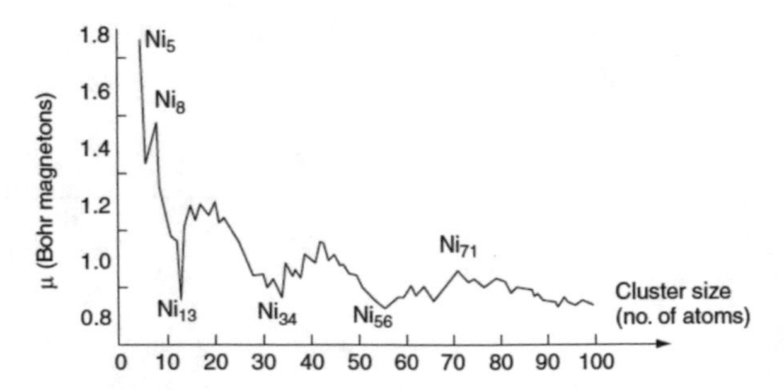

Fig. 2.6 Schematic diagram of the variation in magnetic moments of clusters as a function of cluster size. Reprinted with permission from © Scientific Reports, 7, Article No. 9894 (2017)

only one domain and exhibit a phenomenon known as superparamagnetic. In this case the overall magnetic coercivity is then lowered: the magnetizations of the various particles are randomly distributed due to thermal fluctuations and only become aligned in the presence of an applied magnetic field.

Giant magnetoresistance (GMR) is a spectacle perceived in nanoscale multilayers consisting of a strong ferromagnet (e.g., Fe, Co) and a weaker magnetic or non-magnetic buffer (e.g., Cr, Cu); it is generally engaged in data storage and sensing. In the absence of a magnetic field the spins in alternating layers are oppositely aligned through antiferromagnetic coupling, which gives maximum scattering from the interlayer interface and hence a high resistance parallel to the layers. In an oriented external magnetic field the spins align with each other and this decreases scattering at the interface and hence resistance of the device [10].

2.2.6 *Optical Properties*

The nanoclusters has effect of reduced dimensionality on electronic structure has the most profound effect on the energies of the highest occupied molecular orbital (HOMO), essentially the valence band, and the lowest unoccupied molecular orbital (LUMO), essentially the conduction band. Optical emission and absorption depend on transitions between these states; semiconductors and metals, in particular, show large changes in optical properties, such as colour, as a function of particle size. Colloidal solutions of gold nanoparticles have a deep red colour which becomes progressively more yellow as the particle size increases; indeed gold colloids have been used as a pigment for stained glass since the seventeenth century. Figure 2.7 shows optical absorption spectra for colloidal gold nanoparticles of varying sizes. Semiconductor nanocrystals in the form of quantum dots show similar size-dependent behavior in the frequency and intensity of light emission as well as modified non-linear optical properties and enhanced gain for certain emission energies or wavelengths. Other

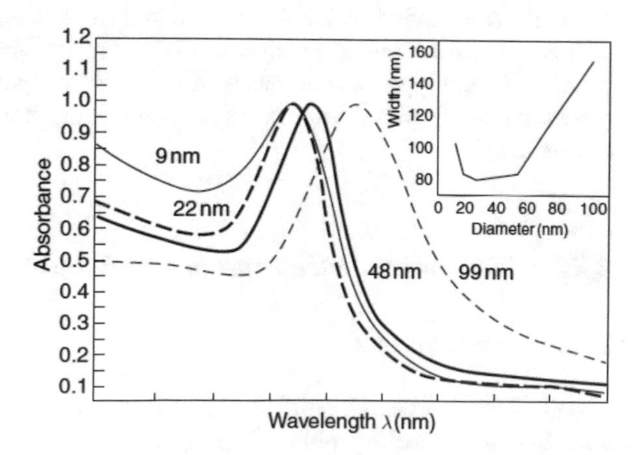

Fig. 2.7 Size dependence of the optical absorption wavelength for gold nanoparticles and (inset) the corresponding value of the full width at half maximum (FWHM) of the absorption peak. Reprinted with permission from © Nanophotonics, 6, 153–175 (2019)

properties which may be affected by reduced dimensionality are photocatalysis, photoconductivity, photoemission and electroluminescence [11].

2.2.7 Electronic Properties

The changes which occur in electronic properties are due to reduction in system length scale and increase in wave-like property of the electrons and the scarcity of scattering centers. As the size of the system becomes similar with the de Broglie wavelength of the electrons, the discrete nature of the energy states becomes apparent once again, although a fully discrete energy spectrum is only observed in systems that are confined in all three dimensions. In certain cases, conducting materials become insulators below a critical length scale, as the energy bands cease to overlap. Owing to their intrinsic wave-like nature, electrons can tunnel quantum mechanically between two closely adjacent nanostructures, and if a voltage is applied between two nanostructures which aligns the discrete energy levels in the density of state (DOS), resonant tunneling occurs, which abruptly increases the tunneling current. In macroscopic systems, electronic transport is determined primarily by scattering with phonons, impurities or other carriers or by scattering at rough interfaces. The path of each electron resembles a random walk, and transport is said to be diffusive. When the system dimensions are smaller than the electron mean free path for inelastic scattering, electrons can travel through the system without randomization of the phase of their wave functions. This gives rise to additional localization phenomena which are specifically related to phase interference. If the system is sufficiently small so that all scattering centers can be eliminated completely, and if the sample boundaries are smooth so that boundary reflections are purely specular, then electron transport becomes purely ballistic, with the sample acting as a waveguide for the electron wavefunction [12]. Conduction in highly confined structures, such as quantum dots, is very sensitive to the presence of other charge carriers and hence the charge state of the dot. These coulomb blockade effects result in conduction processes involving single electrons and as a result they require only a small amount of energy to operate a switch, transistor or memory element. All these phenomena are used to produce completely different types of components for electronic, optoelectronic and information processing applications, such as resonant tunneling transistors and single-electron transistors [8].

2.2.8 Physiochemical Properties of Nanomaterials

2.2.8.1 Melting Point

The number of surface atoms to that of bulk atoms increases with decrease in particle size. Hence, the melting point of nanoparticle falls as much as equivalent surface energy. Some examples of melting points of nanoparticles are shown in Figs. 2.8 and 2.9. A melting point of Au declines suddenly when a particle size is less than 15 nm and a decrease of almost 200 K can be observed at about 6 nm. As shown in Fig. 2.9, it has been reported that there is a linear relationship between a ratio of a melting point of nanoparticle (T_m) to that of bulk material (T_o) and a reciprocal number of a particle diameter.

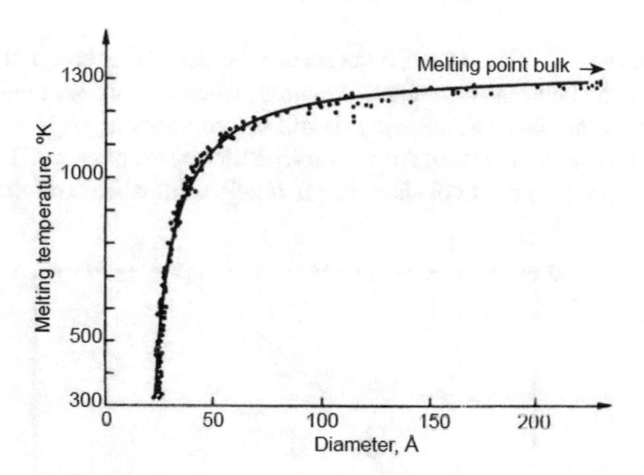

Fig. 2.8 Relationship between melting point of gold and its particle size. Reprinted with permission from © Handbook of Nanoparticles. Springer, Cham (2016)

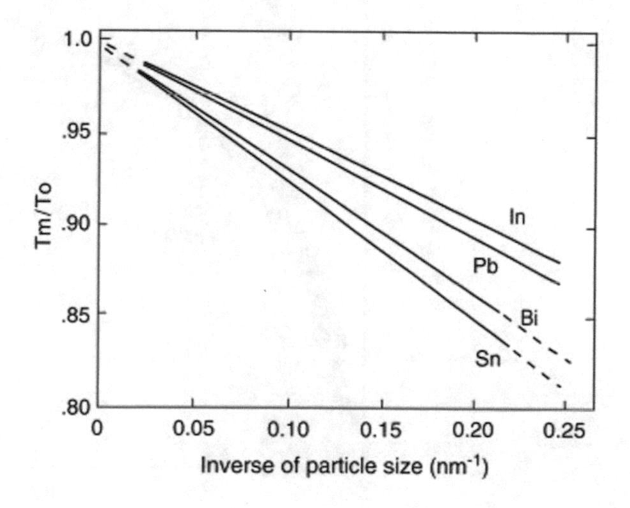

Fig. 2.9 Relationship between inverse of particle size of various metal and T_m/T_o. T_m, measured melting point; T_o, melting point of bulk material (Elsevier Nanoparticle Technology Handbook, Third Edition 2018, pages 3–47)

2.2.8.2 Surface Tension

Researchers calculated a change in liquid drop diameter with an evaporation of a liquid drop at a constant temperature based on a molecular dynamics and obtained an equation

$$\ln(dr/dt) = \ln \mathrm{A} + \mathrm{B}(\sigma/r)$$

where r is the radius of liquid drop, σ the surface tension, t the time, and A and B are constants. As can be seen from above equation, when the surface tension value of liquid drop remains constant, ln(dr/dt) should be proportional to (1/r). Figure 2.10, shows changes of particle sizes of Au and Pb with time by evaporation. It can be found that ln(dr/dt) and (1/r) does not show linear relationship when particle sizes of Au

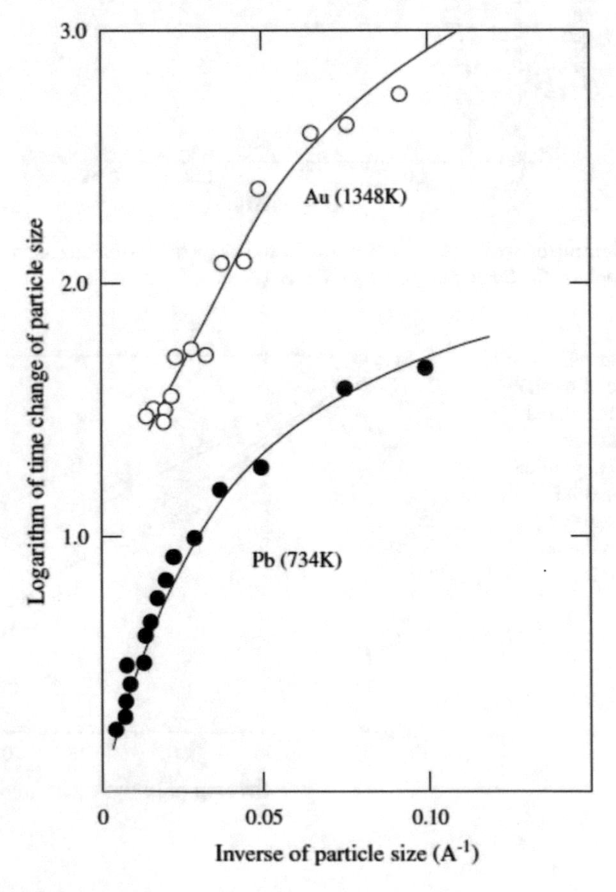

Fig. 2.10 Relationship between particle size change of evaporating metal drop and particle size. Reprinted with permission from © Elsevier Nanoparticle Technology Handbook, 3rd (eds.), 3–47 (2018)

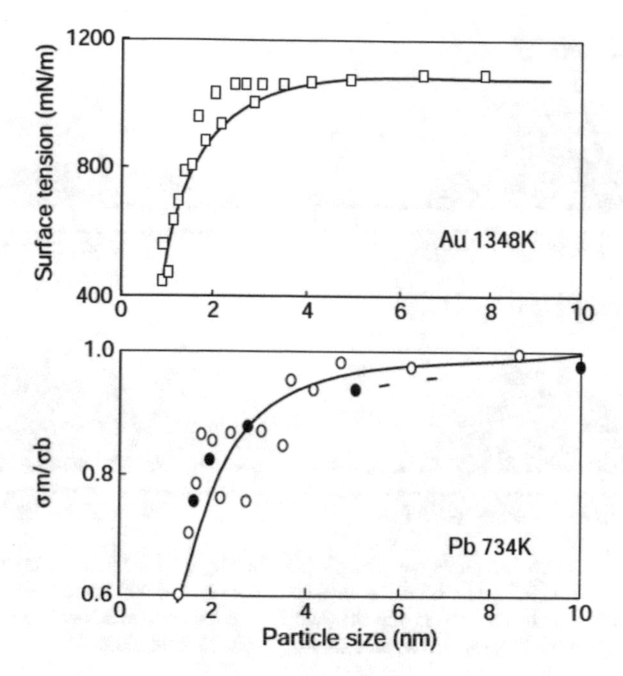

Fig. 2.11 Effect of particle size of surface tension of Au and Pb. σ_m, surface tension of nanoparticle; σ_b, surface tension of bulk metal. Reprinted with permission from © Elsevier Nanoparticle Technology Handbook, 3rd (eds.), 3–47 (2018)

and Pb drops are less than 2 and 5 nm, respectively. The theoretical and experimental approach on surface energies of nanoparticles of metals reports the results of Au and Pb particles [13] (Fig. 2.11).

2.2.8.3 Wettability

As shown in Fig. 2.12, a contact angle value is used as a criterion for wettability of solid by liquid. Generally, the contact angle is larger than 90°, is known as non-wetting system and if the contact angle is smaller than 90°, is known as wetting system. The contact angle of a liquid drop as the result of the equilibrium of mechanical energy between the drop and a solid surface under the action of three interfacial free energies,

$$\sigma_L \cos\theta = \sigma_S - \sigma_{SL}$$

where θ is the contact angle, σ_L the surface free energy of liquid, σ_S the surface free energy of solid and σ_{SL} the interface free energy between solid and liquid. Since the surface free energy of liquid depends on a liquid drop size as mentioned above, the contact angle also depends on the liquid drop size. Figure 2.13 shows particle size dependence of contact angles of various liquid metals on carbon substrate. It can be

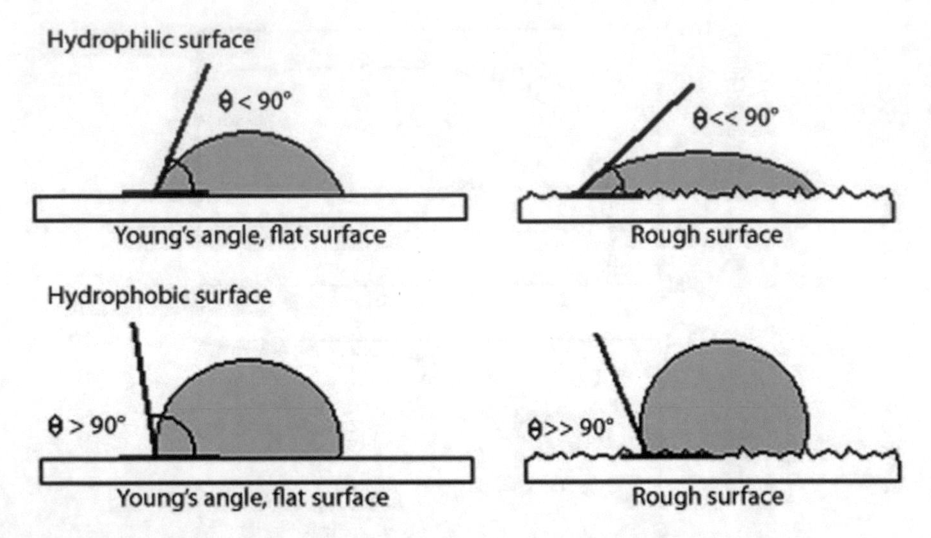

Fig. 2.12 Changes in contact angle with roughness. On a hydrophilic surface (above), the contact angle observed is $\theta \ll 90°$ while on a flat surface it is only $\theta < 90°$. On a hydrophobic surface (below), the contact angle observed is $\theta \gg 90$∘, while on a flat surface it is only $\theta > 90°$. Reprinted with permission from © Progress in Organic Coatings, 65, 77–83 (2009)

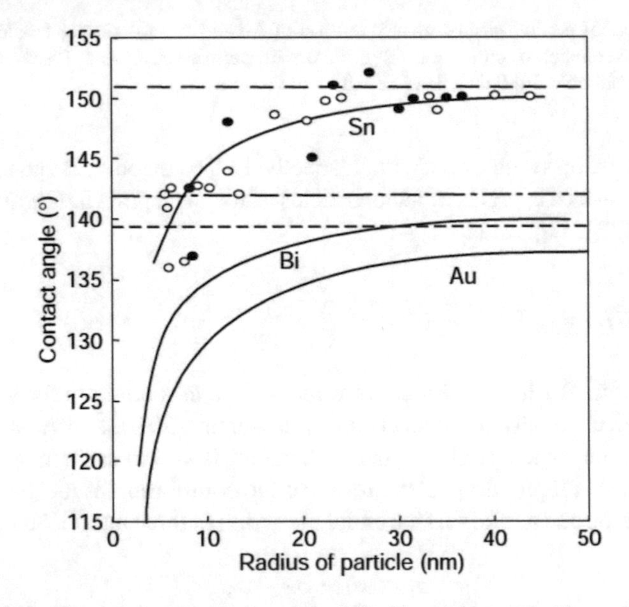

Fig. 2.13 Relationship between contact angles of liquid metals on graphite and particle sizes. Reprinted with permission from © Elsevier Nanoparticle Technology Handbook, 3rd (eds.), 3–47 (2018)

found that the contact angles of all liquid metals decrease at less than 40 nm and a remarkable decrease of the contact angles is observed when particle radius is less than 10 nm [14].

2.2.9 *Specific Surface Area and Pore*

Nanoparticles have large specific surface area, which enhance their surface properties than bulk materials. The specific surface area is boost for the properties of nanoparticles. It is significant to access the effect of particle shapes and particle size distribution for measuring particle size from specific surface area. By using electron microscopic observations or other particle size measurements are preferable in combination with the measurement of specific surface area. Comprehensive analysis for these measurements would enable to cstimate reasonable and meaningful particle size [15].

2.3 Nanomaterial Synthesis Process

Atoms and molecules are vital building blocks of every object. The way in which things are constructed with these basic units is extremely important to understand their properties and their reciprocal interactions. An efficient control of the synthetic pathways is essential during the preparation of nanobuilding blocks with different sizes and shapes that can lead to the creation of new devices and technologies with improved performances. By the way we have two opposite, but complementary approaches are followed. One is a top-down strategy of miniaturizing current components and materials, while the other is a bottom-up strategy of building ever-more-complex molecular structures atom by atom or molecule by molecule. These two different methods highlight the organization level of nanosystems as the crossing point hanging between the worlds of molecular objects and bulk materials.

The top-down approach has been advanced by Richard Feynman in his often-cited 1959 lecture stating that “there is plenty of room at the bottom” and it is ideal for obtaining structures with long-range order and for making connections with macroscopic world. Conversely, the bottom-up approach is best suited for assembly and establishing short-range order at the nanoscale [16].

2.3.1 *Top-Down Approach*

The top-down method is based on miniaturizing techniques, such as machining, templating or lithographic techniques. Top-down methods usually start from patterns generated at larger scale and then they are reduced to nanoscale. A key advantage of the top-down approach is that the parts are both patterned and built in place, so that no further assembly steps are needed. By means of electronic, ionic or X-ray lithography, a monolith can be cut step by step in order to generate a quantum well at first, then a quantum, and a quantum. Recent short-wavelength optical lithography methods can reach dimensions not less than 100 nm. Extreme ultraviolet and X-ray sources are being developed to allow lithographic printing techniques to reach dimensions from 10 to 100 nm, but the principal limits are still due to the difficulty of beam focalization. Likewise, scanning beam techniques such as electron-beam lithography provide patterns down to about 20 nm and still-smaller features are obtained by using scanning probes to deposit or remove thin layers [17].

The general procedure of mechanical printing techniques consist on making a master stamp by a high-resolution lithographic technique, as described above, and then applying this stamp, or subsequent copies of it, to a surface to mould the pattern. The last step is to remove the thin layer of the masking material under the stamped regions. These nanoscale printing techniques offer several advantages due to the possibility to use a wide variety of materials with curved surfaces. As a general drawback, these techniques are not cheap and require a complex manufacturing. In addition, top-down methodologies: (1), even if they work well at the microscale, they collide with some difficulties at nanoscale dimensions and (2) they usually lead to the formation of bidimensional structures and hence, as they are carried out by the addition or subtraction of patterned layers, they cannot easily give rise to the production of arbitrary three-dimensional objects. It is worth underlining that the development in the top-down methodology was mainly driven from the traditional disciplines of materials engineering and physics, whereas the role of inorganic chemists has been minor in the exploitation of these techniques.

2.3.2 Bottom-Up Approach

Bottom-up approaches to nanofabrication involve gradual additions of atoms or groups of atoms. This technique uses chemical or physical forces operating at the nanoscale to assemble basic units into larger structures. The chemical growth of nanometer-sized materials often implies colloidal or supramolecular systems and it frequently passes through phase transformations, such as vapor deposition on surfaces or precipitation of a solid phase from solution. Inspiration for bottom-up approaches comes from biological systems, where nature has employed chemical forces to create essentially all the structures needed by life. Researchers try to mimic nature's ability to produce small clusters of specific atoms, which can then self-assemble into more-elaborated structures [18].

2.4 Applications of Nanomaterials

2.4.1 Environmental Sector

Nanotechnology offers a low-cost and effective solution to the challenge of access to clean and safe water. The technology holds the potential to radically reduce the number of steps, materials and energy needed to purify water. Depending on the kind of water to be purified ground, surface or waste water; nanomaterials can be tailor-made with specific pore sizes and large enhanced surface areas to filter out certain unwanted pollutants, such as heavy metals or biological toxins. For example, titanium oxide at nanoscale can be used to degrade organic pollutants. And silver nano-particles have the ability to degrade biological pollutants such as bacteria. Also, the testing are made with different kinds of membranes and filters based on carbon nanotubes, nanoporous ceramics, magnetic nanoparticles and other nanomaterials which could be used to remove water-borne diseases such as typhoid and cholera, as well as toxic metal ions, organic and inorganic solutes. Another, important concern is nanofiltration, were nanofiltration membranes are already being applied for removal of dissolved salts from salty water, removal of micro pollutants, water softening, and wastewater treatment. These membranes selectively reject substances, which enables the removal of harmful pollutants and retention of nutrients present in water that are required for the normal functioning of the body. Attapulgite clay, zeolite, and polymer filters are source materials for nanofilters and can now are manipulated on the nanoscale to allow for greater control over pore size of filter membranes [19].

Using catalytic particles could chemically degrade pollutants instead of simply moving them somewhere else, including pollutants for which existing technologies are inefficient or cost prohibitive. Magnetic nanoparticles, when coated with different compounds could be used to remove pollutants, including arsenic, from water. The Self-cleaning process includes photo-catalysis and hydrophobicity. These two processes are used in coating of titanium dioxide on the outside surface of the glass Titanium dioxide is an inorganic pigment which is widely used in a whole variety of products and in this case is a very thin coating (25 nm) on the outside surface of the glass [20].

2.4.2 Health Sector

Nanotechnology applications are in development and that will radically improve medical imaging techniques. For example, gold and silver nanoparticles have optical properties which make them extremely effective as contrast agents. Quantum dots which are brighter than organic dyes and need only one light source for excitation, when used in conjunction with magnetic resonance imaging, can produce exceptional images of tumour sites. Nanomaterials are also used in therapeutics or treatment.

2.4.2.1 Targeted Drug Delivery Systems

Nanostructures can be used to recognize diseased cells and to deliver drugs to the affected areas to combat cancerous tumour, for example, without harming healthy cells. In obesity, nanoparticles can target and inhibit the growth of fat deposits.

2.4.2.2 Drug Therapy

Research shows that nano-sized biodegradable polymer capsules containing drugs for tuberculosis treatment are effectively taken up by the body's cells. The effect is a slower release of the drug into the body and a reduction in the frequency with which (Tuberculosis) TB patients need to take his or her medication.

2.4.2.3 Photothermal and Hypothermal Destruction

Nanoparticles like gold (Au) possess therapeutic properties based on their magnetic wavelength or optical properties. They absorb light and heat up the surrounding area, killing the cancer cells. Silver has been used for its ability to destroy bacteria from ancient Romans treating their water with silver coins to NASA using the metal to purify water aboard the Space Shuttle. Silver (Ag) nanoparticles are embedded in sticking plasters for their ability to inhibit the transmission of viruses. Pancreatic cancer has a devastatingly low survival rate because it is usually diagnosed at an advanced stage. Scientists have created tools for the early diagnosis of pancreatic cancer by attaching a molecule that binds specifically to pancreatic cancer cells to iron oxide nanoparticles that are clearly visible under magnetic resonance imaging (MRI). The drug is encapsulated in a nanoparticle which helps it pass through the stomach to deliver the drug into the bloodstream.

2.4.3 Energy Sector

Research based on nanotechnology deals a practical substitute to non-renewable fossil-fuel consumption and provides a chance to realize a hydrogen economy. Nano applications in the above field contain: solar cells; fuel cells and new energy production, conversion and storage processes. In all areas, the results are energy that is cheaper, cleaner, more efficient and renewable. In future, nano holds the potential to produce hybrid vehicles with reduced fuel consumption and a lighter motor weight. Using nanoparticles in the manufacture of solar cells is beneficial due to their reduce manufacturing costs by using a low temperature process instead of the high temperature vacuum deposition process typically used to produce conventional cells made with crystalline semiconductor material. They can reduce installation costs by producing flexible rolls instead of rigid crystalline panels. Currently available nanotechnology solar cells are not as efficient as traditional; however their lower cost offsets this. In the long term nanotechnology versions should both be lower cost and, using quantum dots, should be able to reach higher efficiency levels than conventional ones [21]. Nanostructured devices have the potential to serve as the basis for next-generation energy systems that make use of densely packed interfaces and thin films. Researchers have developed metal-insulator-metal nanocapacitors. It is possible to accommodate one million such tiny capacitors on one square centimeter area. The use of such capacitors in battery and other energy storage devices may increase the efficiency and capacity of such devices enormously [22].

References

1. Issa B, Obaidat IM, Albiss BA et al (2013) Magnetic nanoparticles: surface effects and properties related to biomedicine applications. Int J Mol Sci 14:21266–21305
2. Sidorenko AS (2017) Physics, chemistry and biology of functional nanostructures III. Beilstein J Nanotechnol 8:590–591
3. Ziemann P, Schimmel T (2012) Physics, chemistry and biology of functional nanostructures. Beilstein J Nanotechnol 3:843–845
4. Lohse S (2013) Nanoparticles are all around us. Sustainable nano—a blog by the Center for Sustainable Nanotechnology
5. Sajanlal PR, Sreeprasad TS, Samal AK et al (2011) Anisotropic nanomaterials: structure, growth, assembly, and functions. Nano Rev 2:5883–5944
6. Tian S, Li YZ, Li MB et al (2015) Structural isomerism in gold nanoparticles revealed by X-ray crystallography. Nat Commun 6:8667–8672
7. Putnam SA, Cahill DG, Braun PV (2006) Thermal conductivity of nanoparticle suspensions. J Appl Phys 99:084308–084314
8. Edelstein AS, Cammarata RC (1996) Nanomaterials: synthesis, properties and applications. Institute of Physics, Materials and Manufacturing Process, Taylor and Francis Group, LLC, London
9. Guo D, Xie G, Luo J (2014) Mechanical properties of nanoparticles: basics and applications. J Phys D Appl Phys 47:013001–013025
10. Yurkov YuG, Fionov AS, Koksharov YuA et al (2007) Electrical and magnetic properties of nanomaterials containing iron or cobalt nanoparticles. Inorg Mater 43:834–844
11. Sghaier T, Liepvre SL, Fiorini C et al (2016) Optical absorption signature of a self-assembled dye monolayer on graphene. Beilstein J Nanotechnol 7:862–868
12. Suresh S (2013) Semiconductor nanomaterials, methods and applications: a review. J Nanosci Nanotechnol 3:62–74
13. Mizoguchi T, Dahmen U (2009) 3D shape and orientation of nanoscale Pb inclusions at grain boundaries in Al observed by TEM and STEM tomography. Philos Mag Lett 89:104–112
14. Yokoyama T, Masuda H, Suzuki M et al (2007) Basic properties and measuring methods of nanoparticles. Elsevier, Netherlands, pp 3–48
15. Glover RD, Miller JM, Hutchison JE (2011) Generation of metal nanoparticles from silver and copper objects: nanoparticle dynamics on surfaces and potential sources of nanoparticles in the environment. ACS Nano 5:8950–8957
16. Psaro R, Guidotti M, Sgobba M (2006) Nanosystems. In: Inorganic and bio-inorganic chemistry. EOLSS Publishers, Oxford, UK
17. Guo Z, Tan L (2009) Fundamentals and applications of nanomaterials. ARTECH HOUSE, MA
18. Tirrell M, Requicha A, Friedlander S et al (2000) Synthesis, assembly, and processing of nanostructures. IWGN Workshop Report, Springer, Dordrecht

19. Filipponi L, Sutherland D (2012) Nanotechnologies: principles, applications, implications and hands-on activities. Publications Office of the European Union, Luxembourg
20. Musee N (2009) CSIR speech to the First SA National Workshop on Nanotechnology Risk Assessment Quoted in The Water Wheel (2008). National Centre for Nano Structured Materials. CSIR, Nanotechnology. Water Development
21. Wan J, Song T, Flox C et al (2015) Advanced nanomaterials for energy related applications. J Nanomater 564097:2
22. Liu N, Li W, Pasta M et al (2014) Nanomaterials for electrochemical energy storage. Front Phys 9:323–350

Chapter 3
Fundamentals of Nanostructures

Abstract Nanostructures may define by its dimensions as one dimension, two dimension and three dimensions based on quantum confinement effects. Nanostructures are made of nanoparticles using atom by atom or electron arrangement with the help of deposition and lithography techniques. This chapter briefly describe about nanostructured materials, approach towards nanostructured materials and various types of nanostructural materials.

3.1 Nanostructures Definition

Nanostructures are related to a human hair which is ~50,000 nm thick whereas the diameters of nanostructures are ~0.3 nm for a water molecule, 1.2 nm for a single-wall carbon nanotube, and 20 nm for a small transistor. DNA molecules are 2.5 nm wide, proteins about 10 nm, and an ATPase biochemical motor about 10 nm. This is the ultimate manufacturing length scale at present with building blocks of atoms, molecules, and supramolecules as well as integration along several length scales. In addition, living systems work at the nanoscale [1]. Nanostructure can be defined as a system in the order of 1–100 nm in size. Also, the nanostructures have intermediate size between a nano dimension and a micro dimension, which can be developed as various forms. It is basically structured using soft/hard templates to form micro-level structures. Nanostructure means it represents the structure of the nanomaterials, like sphere, rod, hollow cylinder, rectangle, cube, etc. nanomaterial means materials in nano size. It represents each and every material present in nano size and it may be biological or non-biological, artificial or natural, anything comes under nano size or nanomaterials. Actually nanostructured materials are a right way to represents a material in nano size. e.g.: nano tubes, nano rods, nanoparticles, nanofibers, nanosheets etc. Nanostructures are the building blocks of nanomaterials [2, 3].

T. D. Thangadurai et al., *Nanostructured Materials*, Engineering Materials,
https://doi.org/10.1007/978-3-030-26145-0_3

3.2 Nanostructured Materials

Nanostructured materials are condensed material, in whole or in part composed of structural elements with the characteristic dimensions from several nanometers to several tens of nanometers; the long-range order in the structural elements is highly disturbed and therefore the short-range order determines multiparticle correlations in the arrangement of atoms in these elements; all macroscopic properties of the material are determined by the size and/or mutual arrangement of structural elements [4]. One way to classify nanostructures is based on the dimensions in which electrons move freely:

Quantum well: electrons are confined in one dimension (1D), free in other 2D. It can be realized by sandwiching a narrow-bandgap semiconductor layer between the wide-gap ones. A quantum well is often called a 2D electronic system. **Quantum wires**: confined in two dimensions, free in 1D. Real quantum wires include polymer chains, nanowires and nanotubes. **Quantum dots**: electrons are confined in all dimensions, as in clusters and nano-crystallites. Zero-dimensional nanomaterials include nanocluster materials and nanodispersions, i.e. materials in which nanoparticles are isolated from each other. One-dimensional nanomaterials are nanofibre (nanorod) and nanotubular materials with fibre (rod, tube) length from 100 nm to tens of microns. Two-dimensional nanomaterials are films (coatings) with nanometer thickness. Structural elements in 0D, 1D and 2D nanomaterials can be distributed in a liquid or solid macroscopic matrix or be applied on a substrate. Three-dimensional nanomaterials include powders, fibrous, multilayer and polycrystalline materials in which the 0D, 1D and 2D structural elements are in close contact with each other and form interfaces. An important type of three-dimensional nanostructured materials is a compact or consolidated (bulk) polycrystal with nanosize grains, whose entire volume is filled with those nanograins, free surface of the grains is practically absent, and there are only grain interfaces. The formation of such interfaces and disappearance of the nanoparticle (nanograin) surface is the fundamental difference between three-dimensional compact nanomaterials and nanocrystalline powders with various degrees of agglomeration that consist of particles of the same size as the compact nanostructured materials [5].

Nanostructured materials made of nanosized grains or nanoparticles as building blocks, have a significant fraction of grain boundaries with a high degree of disorder of atoms along the grain boundaries, and a large ratio of interface area to volume. Chemical composition of the phases and the interfaces, between nano-grains, must be controlled as well. One of the most important characteristics of nanostructured materials is the dependence of certain properties upon the size in nanoscale region. For example, electronic property, with quantum size effects, caused by spatial confinement of delocalized valence electrons, is directly dependent on the particle size. Small particle size permits conventional restrictions of phase equilibrium and kinetics to be overcome during the synthesis and processing by the combination of short diffusion distances and high driving forces of available large surfaces and interfaces. A wide range of materials, including metals and ceramics in crystalline, quasi-crystalline, or amorphous phases have been synthesized as nanosized or nanostructured materials. The large surface area gives higher reactivity. Thus novel properties may result from surface defects. In addition, there are other structural features in the nanostructured materials that depend on the manner in which these materials are synthesized and processed, such as surface pores, grain boundary junctions, and other crystal lattice defects. Some examples are lowered melting temperatures, improved wear resistance of nanostructured ceramics, increased strength of metals, and transformation of magnetic state from ferromagnetic to paramagnetic or superparamagnetic state as a function of size. The changes in lattice parameter can be attributed to changes in surface stress, while the reduction in melting temperature results from the increase in surface free energy. This opens the doors for tailoring given properties by careful synthesis of the building blocks and their assembly to fabricate functional materials with improved properties [6].

Ultrafine microstructures having an average phase or grain size on the order of a nanometer (10^{-9} m) are classified as nanostructured materials (NSMs). The interest in these materials has been stimulated by the fact that, owing to the small size of the building blocks (particle, grain, or phase) and the high surface-to-volume ratio, these materials are expected to demonstrate unique mechanical, optical, electronic, and magnetic properties. The properties of NSMs depend on the following four common microstructural features: (1) fine grain size and size distribution (<100 nm); (2) the chemical composition of the constituent phases; (3) the presence of interfaces, more specifically, grain boundaries, heterophase interfaces, or the free surface; and (4) interactions between the constituent domains. In nanophase materials, a variety of size-related effects can be incorporated by controlling the sizes of the constituent components. For example, nanostructured metals and ceramics can have improved mechanical properties compared to conventional materials as a result of the ultrafine microstructure. In addition, NSMs have the capability to be sintered at much lower temperatures than conventional powders, enabling the full densification of these materials at relatively lower temperatures [7]. Semiconductor NSMs are currently also considered to have technological applications in optoelectronic devices such as semiconductor quantum dots and photodiodes, owing to the quantum size effects caused by the spatial confinement of delocalized electrons in confined grain sizes. Magnetic applications of NSMs include fabrication of devices with giant magnetoresistance (GMR) effects, the property used by magnetic heads to read data on computer hard drives, as well as the development of magnetic refrigerators that use solid magnets as refrigerants rather than compressed ozone-destroying chlorofluorocarbons. In addition, nanostructured metals and ceramics seem to be candidates for new catalytic applications [8]. The development of semiconductor nanoclusters is an area of intense research efforts. These nanoclusters are often referred to as quantum dots, nanocrystals, and Q-particles. In the nanometer size regime, electron-hole confinement in nanosized spherical semiconductor particles results in three-dimensional size quantization. Band gap engineering by size and dimension quantization is important because it leads to electrical, optical, magnetic, optoelectronic and magneto optical properties substantially different from those observed for the bulk material. As an example, quantum dots can be developed to emit and absorb a desired wavelength of light by changing the particle diameters [9].

3.2.1 1D Nanostructures

One-dimensional (1D) nanostructures are ideal systems for investigating the dependence of electrical transport, optical properties and mechanical properties on size and dimensionality. They are expected to play an important role as both interconnects and functional components in the fabrication of nanoscale electronic and optoelectronic devices. Nanoscale one-dimensional (1D) materials have stimulated great interest due to their importance in basic scientific research and potential technology applications. Other than carbon nanotubes, 1D nanostructures (nanowires or quantum wires) are ideal systems for investigating the dependence of electrical transport and mechanical properties on size and dimensionality. They are expected to play an important role as both interconnects and functional components in the fabrication of nanoscale electronic and optoelectronic devices. Many unique and fascinating properties have already been proposed or demonstrated for this class of materials, such as superior mechanic toughness, higher luminescence efficiency, enhancement of thermoelectric figure of merit and lowered lasing threshold. Nanowires are anisotropic nanocrystals with large aspect ratios. Generally, they would have diameters of 1–200 nm and length up to several tens of micrometers. Nanowires differ significantly from spherical nanocrystals by their morphology as well as physical properties. An important issue in the study and application of these 1D materials is how to assemble individual atoms into 1D nanostructure in an effective and controllable way. Although 1D nanostructures could be fabricated using a number of advanced nanolithography techniques, such as e-beam writing, proximal-probe patterning, and x-ray lithography, these processes, however, generally are slow and the cost is high, the development of these techniques into practical routes for fabricating large numbers of 1D nanostructures rapidly and at low-cost still requires great ingenuity [10].

3.2.2 2D Nanostructures

2D nanostructures have two dimensions outside of the nanometric size range. In recent years, a synthesis 2D NSMs have become a focal area in materials research, owing to their many low dimensional characteristics different from the bulk properties. In the quest of 2D NSMs, considerable research attention has been focused over the past few years on the development of 2D NSMs. 2D NSMs with certain geometries exhibit unique shape-dependent characteristics and subsequent utilization as building blocks for the key components of nanodevices. In addition, a 2D NSMs are particularly interesting not only for basic understanding of the mechanism of nanostructure growth, but also for investigation and developing novel applications in sensors, photocatalysts, nanocontainers, nanoreactors, and templates for 2D structures of other materials. The 2D NSMs, are junctions (continuous islands), branched structures, nanoprisms, nanoplates, nanosheets, nanowalls, and nanodisks [11].

3.2.3 3D Nanostructures

Owing to the large specific surface area and other superior properties over their bulk counterparts arising from quantum size effect, 3D NSMs have attracted considerable research interest and many 3D NSMs have been synthesized in the past years. It is well known that the behaviors of NSMs strongly depend on the sizes, shapes, dimensionality and morphologies, which are thus the key factors to their ultimate performance and applications. Therefore it is of great interest to synthesize 3D NSMs with a controlled structure and morphology. In addition, 3D nanostructures are an important material due to its wide range of applications in the area of catalysis, magnetic material and electrode material for batteries. Moreover, the 3D NSMs have recently attracted intensive research interests because the nanostructures have higher surface area and supply enough absorption sites for all involved molecules in a small space. The typical 3D NMSs are nanoballs, nanocoils, nanocones, nanopillers and nanoflowers [12].

3.3 Features of Nanostructures

Nanostructures are unique as compared with both individual atoms/molecules at a smaller scale and the macroscopic bulk materials. They are also called mesoscopic structures. Nanoscience research focuses on the unique properties of nanoscale structures and materials that do not exist in structures of same material composition but at other scale ranges. By controlling the construction of the materials on the atomic level, new and improved mechanical, chemical and optical properties may be developed. Both surface structure and particle size on the nanoscale are of great importance in this context. Developments of new theories and models, as well as characterization methods are important in order to understand the relationship between structure and function. Nanostructure materials are often in a metastable state. Their detailed atomic configuration depends sensitively on the kinetic processes in which they are fabricated. Therefore, the properties of nanostructures can be widely adjustable by changing their size, shape and processing conditions [13]. Nanoscale materials offer larger surface area (A) to volume (V) ratios (A/V) than the bulks. Thus nanostructures are believed to have better performances than materials in the micro- or larger scales for sensing applications. The most example for nanostructures are Zinc Oxide, which can appear as nanowires, nanobelts, nanopropellers, nanocombs, nanotubes, nanoswords, nanotripods, nanotetrapods, nanosquids, nanorods, nanotips, etc.

3.4 Theoretical Substantiation of the Approaches Proposed

The activity of nanostructures in self-organization processes is defined by their surface energy thus corresponding to the energy of their interaction with the surroundings. It is known that when the size of particles decreases, their surface energy and particle activity increase. The following ratio is proposed to evaluate their activity:

$$a = \varepsilon_S / \varepsilon_V \quad (3.1)$$

where, ε_S—nanoparticle surface energy, ε_V—nanoparticle volume energy. Naturally in this case $\varepsilon_S \gg \varepsilon_V$ conditioned by the greater surface "defectiveness" in comparison with nanoparticle volume. To reveal the dependence of activity upon the size and shape we take ε_S as $\varepsilon_S^0 S$, and $\varepsilon_V = \varepsilon_V^0 V$, where ε_S^0—average energy of surface unit, S—surface, ε_V^0—average energy of volume unit, V—volume, then Eq. (3.1) is converted to:

$$a = d \, . \frac{\varepsilon_S^0}{\varepsilon_V^0} S / V \quad (3.2)$$

Substituting the values of S and V for different shapes of nanostructures, we see that in general form the ratio S/V is the ratio of the number whose value is defined by the nanostructure shape to the linear size connected with the nanostructure radius or thickness. Equation (3.2) can be given as,

$$a = d \, . \frac{\varepsilon_S^0}{\varepsilon_V^0} N / r(h) = \frac{\varepsilon_S^0}{\varepsilon_V^0} 1 / B \quad (3.3)$$

where, B equals r(h)/N, r—radius of bodies of revolution including hollow ones, h—film thickness depending upon its "distortion from plane", N—number varying depending upon the nanostructure shape. Parameter d characterizes the nanostructure surface layer thickness, and corresponding energies of surface unit and volume unit are defined by the nanostructure composition. For the corresponding bodies of revolution the parameter B represents an effective value of the interval of nanostructure linear size influencing the activity at the given interval r from 1 to 1000 nm. For nanofilms the surface and volume are determined by the defectiveness and shape of changes in conformations of film nanostructures depending upon its crystallinity degree. However, the possibilities of changes in nanofilm shapes at the changes in the medium activity are higher in comparison with nanostructures already formed. At the same time, the sizes of nanofilms formed and their defectiveness play an important role [14].

To determine the possibility of nanostructure existence and forecasting the formation of different nanosystems it is necessary to develop the corresponding theoretical and computation apparatus. The creation of such an apparatus is possible based on corresponding ideas of quantum chemistry, molecular mechanics and thermodynamics. Apparently, in developing the overall theory of nanosystems the symbiosis of similar theoretical trends in physics, chemistry, biology, and computer modeling will be found. The processes of nanostructure formation, the subject investigated by chemical physics, are of great interest. Therefore, apparently, it is more appropriate to discuss the creation of computational apparatus of chemical physics of nanostructure formation processes. In recent years, a lot of investigations connected with the development of different software products have been carried out in quantum chemistry, molecular dynamics and molecular mechanics of cluster and nanosystems. These investigations are fulfilled using semi-empirical and ab initio methods. Basically, paired interactions were used in all the methods, although there are works discussing collective interactions on atomic and molecular levels [15]. It is advisable to arrange a hierarchical scheme for predicting nanostructure formation. First, the stability of interacting particles and reaction centers should be determined. Under reaction center we understand the group of atoms being changed during the reaction. Potential reactivity centers, that is, groups of atoms able to participate in reactions, are usually called functional groups. This terminology can mainly be found in organic and polymer chemistry. The stability of particles and interaction energy are evaluated with methods of quantum chemistry. After reactivity centers are selected possible reactions with their participation are assumed, this operation for finding reaction options is similar to the isolation of reaction series in physical organic chemistry. Here, under reaction series we understand the directivity of processes in reactivity centers or similar nanosystems with the formation of nanoparticles of certain structure and composition.

The advantage of one or another reaction series when compared with the others is determined based on energy consumption for performing successive acts of the process and the rate with which these acts are performed [16]. Computational experiment can be carried out using apparatuses of quantum chemistry and molecular dynamics, in some cases in combination with semi-empirical methods of thermodynamics in the frameworks of activated complex theory or spatial energy concept. At previous two stages of nanosystems modeling and formation single chemical particles, reaction centers or fragments of complex nano components of the systems were discussed. The changes of nanostructures during the action of various fields, such as thermal, electrical, magnetic, field of particles, gravitation upon nanosystems, and nanoparticles are also determined at the aforesaid stages. Therefore, at third and the following stages collective interactions of nanoparticles with the help of apparatuses of molecular mechanics and thermodynamics are discussed. The transitions from one method of computational experiment to another one represent certain terminological and conceptual difficulties. This can be explained not only by the time frames when the corresponding notions, ideas and definitions appeared [17]. The complete positive material accumulated in reactivity theory and reactions of chemical particles with the formation of nanosystems should be used in a new nanoscience. Therefore, modular construction should be introduced into the scheme of basic notions of nanosystems and nanostructure modeling together with hierarchical structure (Fig. 3.1).

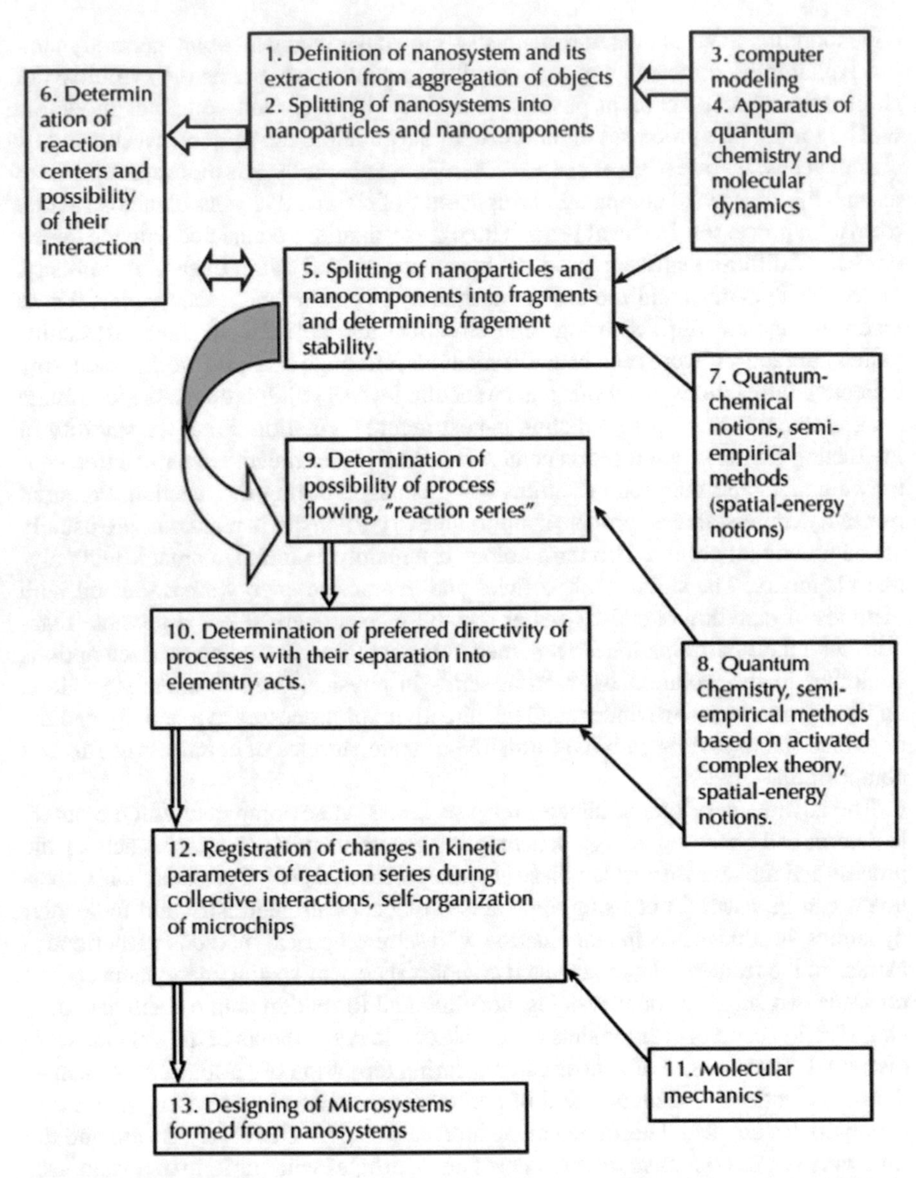

Fig. 3.1 Scheme of nanosystems consideration. Reprinted with permission from © CRC, Taylor & Francis Group (2014)

3.5 Types of Nanostructured Material

The type of nanostructures includes nanotubes, fullerenes, nanowires, nanorods, nanoonions, and nanobulbs. Thin films with thicknesses below 100 nm, which itself can consist of nanomaterials, and Nanocomposites (Table 3.1).

Table 3.1 Nanostructured materials by size dependence

Nanostructure	Size	Material
Clusters, nanocrystals quantum dots	Radius, 1–10 nm	Insulators, semiconductors, metals, magnetic materials
Other nanoparticles	Radius, 1–100 nm	Ceramic oxides
Nanobiomaterials, photosynthetic reaction	Radius, 5–10 nm	Membrane protein
Nanowires	Diameter, 1–100 nm	Metals, semiconductors, oxides, sulfides, nitrides
Nanotubes	Diameter, 1–100 nm	Carbon, layered chalcogenides, BN, GaN
Nanobiorods	Diameter, 5 nm	DNA
Two-dimensional arrays of nanoparticles	Area, several nm^2—μm^2	Metals, semiconductors, magnetic materials
Surfaces and thin films	Thickness, 1–100 nm	Insulators, semiconductors, metals, DNA
Three-dimensional superlattices of nanoparticles	Several nm in three dimensions	Metals, semiconductors, magnetic materials

The nanoparticle dimensions are determined by its formation conditions. When the energy consumed for macroparticle destruction or dispersion over the surface increases, the dimensions of nanomaterials are more likely to decrease. Several researchers consider nanomaterials to be aggregations of nanocrystals, nanotubes, or fullerenes. Simultaneously, there is a lot of information available that nanomaterials can represent materials containing various nanostructures. Some of the nanostructures are fullerenes, gigantic fullerenes, fullerenes filled with metal ions, fullerenes containing metallic nucleus and carbon shell, one-layer nanotubes, multi-layer nanotubes, fullerene tubules, scrolls, conic nanotubes, metal-containing tubules, onions, Russian dolls, bamboo-like tubules, beads, welded nanotubes, bunches of nanotubes, nanowires, nanofibers, nanoropes, nano semispheres, nanobands and various derivatives from enlisted structures [18, 19]. Nanoflowers, these small plants can only be seen properly through an electron microscope. Actually flower shape crystals rather than actual flowers; they build themselves one molecule at a time using an underwater chemical reaction (Figs 3.2 and 3.3).

3.5.1 *Nanostructures in Plants*

Plant surfaces like leaves; contain nanostructures that are used for abundant resolves such as insects sliding, mechanical stability, and increased visible light and harmful UV reflection and radiation absorption. The well-known nanostructure property in plants is the super hydrophobicity in lotus leaves that helps in self-cleaning and super-wettability of the leaves. Many studies in the literature have suggested that stacks of

Fig. 3.2 **a** Nanoflower—p153 'nanoflower' grown from chemical solution and p44 'nanoflowers' created using carbon dioxide gas, **b** Nanoflower—p152 Nanoflowers grown from chemical solution, **c** Nanospheres, and **d** Nanoflakes. Reprinted with permission from © NanoScience: Giants of the Infinitesimal by Peter Forbes and Tom Grimsey and published by Papadakis

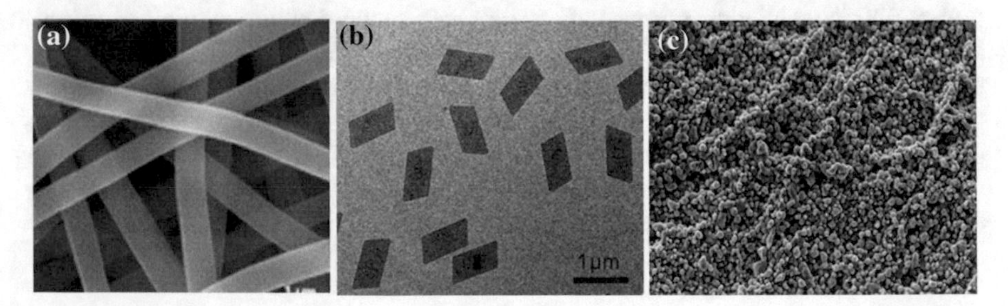

Fig. 3.3 **a** Nanofibers, **b** Nanosheets, and **c** Close-up of the nanostructures grown on cotton textiles. Reprinted with permission from © RMIT University researchers

Fig. 3.4 **a** Photograph of a lotus leaf; **b** SEM image of the lotus leaf surface. The inset is a SEM image of typical 5–9 μm micropapillae covering the surface with fine branch-like nanostructures, copyright 2002 Wiley-VCH. **c** Photograph of a red rose and **d** SEM image of a rose petal surface. The inset is a magnified SEM image of the microcapillary arrays. Reprinted with permission from © American Chemical Society (2008)

nanostructures are responsible for the circular layer in plants and insects which allows them to float on water without sinking. Based on these reports, many artificial super hydrophobic materials with self-cleaning ability have been manufactured through electrodeposition, photolithography and colloidal systems with unique morphology and roughness. These super hydrophobic materials were useful in applications such as water treatment, wettability switchers, smart actuators, transparent coatings and electrodes [20] (Fig. 3.4).

3.5.2 Nanostructures in Insects

Insect wing membranes are comprised of building materials with 0.5 μm–1 mm thickness. Additionally, the insect wings are formed by a complex vein system which gives superior stability to the entire wing structure. Long chain crystalline chitin polymer is the basic framework of insect wings that provides membrane support and allows for bearing forces on them during flight. Resilin enhances the wing's flexibility and is a unique component that is found in between the junctions of the vein and the wing. The routine and longer colonization flights were supported by the vein system along with their weightless wing material. Insect wing surfaces demonstrate a rough and highly ordered structure comprised of micro- and nanoscale properties to minimize their mass and protect them against wetting and pollutants.

Animals such as flies, spiders, and geckos with varying body weight can attach along ceilings and move along vertical walls. The interaction of their patterned surface structure with the substrate profile gives efficient ability and mechanism for attachment to the insect's legs. An intense inverse scaling effect in these attachment devices are exposed via an extensive microscopic study. It has been shown that adhesion is ensured by sub-micrometric devices whereas flies and beetles rely on terminal setae that are of micrometer dimensions. The principle of contact mechanics, which shows that the adhesion leads to the splitting of contacts into finer subcontacts, helps to clearly explain the insect body weight to set a trend. The natural adhesive system uses this principle for their design and may be incorporated in future practical applications. Electrostatic forces, sticking fluids, and microsuckers are the proposed reasons that explain the insect's attachment mechanism. Some of these theories have been rejected based on experimental data and combination of secretion-mediated capillary attractive forces and molecular interactions or van der Waals interactions leads to adhesion. This may be due to the production of secretory fluids in the contact area by some animals, which makes the basic force in the physical form contribute to their adhesion. The reason for adhesion of gecko setae is due to van der Waals interaction through strong evidence and rejects the capillary adhesion mechanisms. It was predicted that application of contact mechanics may help in smaller set an array endings by releasing greater adhesive strength [21].

3.5.3 Nanostructures in the Human Body

The human body consists of nanostructures without which normal function of the body is impossible. It is formed by nanostructures such as bones, enzymes, proteins, antibodies and DNA. Even some works categorize bone as a nanomaterial comprised of hierarchical inorganic nano hydroxyapatite and organic collagen. Additionally, microorganisms such as viruses and bacteria are nanostructures that can cause diseases in humans [2, 3].

3.5.3.1 DNA Nanostructures

DNA is the genetic blueprint repository of living organisms. It helps in the synthesis of protein, which is essential for the activities of living organisms. Monophosphorylated deoxyribose sugar attached with nitrogenized aromatic nucleobase is called a nucleotide, and this is the basic structural unit of DNA. DNA possesses diverse sequence information storage mechanisms with 2.86 bits per linear nanometer density. A-DNA, B-DNA, and Z-DNA are three types of DNA classification based on the base-paring between the strands. B-DNA is a right handed double-helical DNA structure whereas A-DNA is a comparatively short, more-compact, right handed double-helical structure, and Z-DNA is a left-handed double-helical DNA formed with long polypurine stretches. These DNAs are nanostructures in organisms and their interactions with other NMs play a major role in nanomaterial drug formulations [22]. Thus, in recent years, research on artificial DNA nanostructures has escalated in the field of bionanotechnology.

A phosphate backbone with negative charge, nucleobase with metal chelates, and the hydrophobic core with aromatic rings are the chemical handles that are responsible for the formation of self-assembled nanostructures through interaction with inorganic NMs. The formation of DNA-templated metal nanostructures is possible by localizing transition metal cations on DNA to act as precursors and chemical handles. DNA nanostructures and DNA attached to NPs have been synthesized for various applications including nano barcoding and DNA sensors. Research in this area has advanced to include active self-reconfiguration of 1, 2 or 3-dimensional DNA-based nanoscale architectures for drug delivery, molecular electronics and logics. Recent developments in DNA technologies such as Holliday junction elucidation and crossovers help in the virtual assemblage of any DNA structures through DNA origami. An extensive review on DNA origami, their functions and potential has been reported in. They mentioned that NP-templated DNA and hybridization- based DNA are revolutionary particles that will create a positive impact on future biomedical fields [23].

3.5.3.2 Protein Nanostructures

Antibodies, enzymes, proteins and most organelles within cells are smaller than the micrometer-scale and are considered nanostructures. Recently, lipids, self-assembled peptides, and polysaccharides were also included in the list of nanostructures present in the human body. These nanostructures are artificially manipulated for use in pharmaceutical industries. Nanozyme, which is an example of such nanostructures, is an engineered nanometer-scaled artificial enzyme. The enzyme functions to mimic the general natural enzyme principles. Cyclodextrins, porphyrins, supramolecules, polymers and biomolecules, which include antibodies, nucleic acids and proteins, have been widely investigated to imitate the structure and function of natural enzymes. Nanozymes are already under research for applications in biosensing, immunoassays, stem cell growth and environmental rehabilitation via pollutant removal. The viral protein capsids are extensively under research investigation as self-assembling NPs. Aside from that, manipulation of natural proteins and antibodies with NPs as well as individual proteins/antibodies are gaining positive biomedical applications. It is believed that these biomolecular NPs will be highly beneficial for efficient biomolecule delivery and in therapies and diagnostics for complex diseases and genetic disorders [2, 3].

3.5.4 Ceramic Nanostructures

Nanostructured ceramics are one class of materials that can successfully be densified by dynamic compaction, but the technique is certainly not limited to these materials. Magnetic pulse compaction is a commercially used technique for near-net-shape densifying metal powders into all kinds of shapes with densities close to the theoretical density [24, 25]. The most widespread application is the shaping of gear wheels for the automotive industry, but also connecting different metals (cold welding) and polymer-metal connections are performed commercially using MPC. The advantage of commercial equipment is the ability to compact one sample per second.

3.5.5 Polymer Nanostructures

Recently, the nanostructured metal-polymer and metal oxide—polymer composites are the subject of increased attention of researchers in different areas of science and technology. Among the perspectives for using these materials are novel types of solar cells and chemical gas sensors. The electrical, mechanical, and optical properties of the nanocomposites are improved by dispersing a metallic phase in a polymeric matrix. Nanostructured metal (Pd, Sn, Cu, Al)-polymer (poly-para-xylylene) and metal-oxide-polymer composites reveal synergism of properties of the initial components, which gives rise to specific electrical, mechanical, optical and chemical properties related to an ordered distribution of nanoparticles over the matrix volume [26]. Various methods of preparing a metal-polymer nanocomposite exist and from the point of view of homogeneity, methods where polymerization and formation of the nanoparticles are performed simultaneously are very promising.

3.5.6 Nanocomposites

Composite materials are engineered materials made from two or more constituent materials with significantly different physical or chemical properties, which remain separate and distinct on a macroscopic level within the finished structure. One of the most successful composites in history is concrete; other well-known examples are plywood and carbon fiber reinforced plastics. By combining two materials to a composite, physical or chemical properties are achieved which are not possible with a single, homogeneous material. It should not be concealed that also cost aspects have played a role in the development of nanocomposites, as one can combine an expensive material of distinct properties with a low cost filler material. Nanocomposites are nanotechnology's logical progression of composite materials. In nanocomposites, at least one of the constituents is of nanometer size. Thus, in a certain sense nanocomposites also include porous materials, gels, co-polymers and colloids. By this approach, a new class of materials with extraordinary optical, magnetic, electrical and mechanical properties can be achieved. Another important aspect of nanocomposites is the fact that—compared to macroscopic composites—the surface to volume ratio is much higher, typically by at least one order of magnitude. As, for example, the mechanical properties of nanocomposites are to a large extent determined by the interface between the two components, this means that much stronger effects can be obtained by nanocomposites, or that much less amounts of the expensive reinforcement material (e.g. nanotubes, nanofibers) are necessary to obtain the same effect. One special group of nanocomposites are nanostructures glasses (glasses are per definition amorphous, non-crystalline solids). In fact, glasses containing nanoparticles are probably the oldest man-made nanocomposites: medieval stained church glasses contain gold or silver nanoparticles with diameters of 50–100 nm. At the present time, nanocomposite and other nanostructured glasses are owing to their outstanding optical, electrical, mechanical and chemical properties again intensively investigated [27].

3.5.7 Thin Films

Mostly, thin films are used to render the surface properties of a bulk material in those cases where a change of the bulk properties is too expensive or simply impossible to achieve for physical or chemical reasons. In addition, combinations of thin films may serve to achieve complex combinations of surface properties. Thin films are two-dimensional nanostructures by definition. In the field of optics, thin films are defined as structures in which interferences effects don't play a role, i.e. they must be on the order of the wavelength of the light or even thinner. In the field of semiconductor electronics, active and passive layers of about 1.2 nm are currently investigated. Here, the films are moreover meanwhile structured laterally to dimensions of a few tens of nanometers. In recent times, besides homogeneous films also nanostructured films play an ever increasing role. Types of films, (i) Multilayer films consisting of stacks of individual films each of which possesses a thickness of a few nm only, (ii) Nanocomposite films consisting of nanocrystals embedded in an amorphous matrix, (iii) Nanocomposite films consisting of nanoparticles, nanotubes etc. embedded in an amorphous matrix [28].

3.5.8 Nanostructure Computation

Nanostructures also offer opportunities for meaningful computer simulation and modeling since their size is sufficiently small to permit considerable rigor in treatment. In computations on nanomaterials, one deals with a spatial scaling from 1 Å to 1 μm and temporal scaling from 1 fs to 1 s, the limit of accuracy going beyond 1 kcal/mol. There are many examples to demonstrate current achievements in this area: familiar ones are STM images of quantum dots and the quantum corral of 48 Fe atoms placed in a circle of 7.3-nm radius. Ordered arrays or superlattices of nanocrystals of metals and semiconductors have been prepared by several workers. Nanostructured polymers formed by the ordered self-assembly of triblock copolymers and nanostructured high-strength materials are other examples. Prototype circuits involving nanoparticles and nanotubes for nanoelectronics devices have been fabricated [29].

References

1. Gleiter H (2000) Nanostructured materials: basic concepts and microstructure. Acta Mater 48:1–29
2. Jeevanandam J, Barhoum A, Chan YS (2018) Review on nanoparticles and nanostructured materials: history, sources, toxicity and regulations. Beilstein J Nanotechnol 9:1050–1074
3. Jeevanandam J, Barhoum A, Chan YS et al (2018) Review on nanoparticles and nanostructured materials: history, sources, toxicity and regulations. Beilstein J Nanotechnol 9:1050–1074
4. Koch CC (2006) Nanostructured materials: processing, properties and applications, 2nd edn. Elsevier Science
5. Gusev AI, Rempel AA (2004) Nanocrystalline materials. Cambridge International Science Publishing
6. Logothetidis S (2012) Nanostructured materials and their applications. Nanoscience and technology. Springer, Berlin, Heidelberg
7. Rempel AA (2007) Nanotechnologies: properties and applications of nanostructured materials. Russ Chem Rev 76:435–461
8. Fafard S, Hinzer K, Allen CN (2004) Semiconductor quantum dot nanostructures and their roles in the future of photonics. Braz J Phys 34:550–554
9. Sumith AM, Nie S (2010) Semiconductor nanocrystals: structure, properties, and band gap engineering. Acc Chem Res 43:190–200
10. Cao H (2018) Synthesis and applications of inorganic nanostructures. Wiley-VCH Verlag GmbH & Co, KGaA
11. Tiwari JN, Tiwari RN, Kim KS (2012) Zero-dimensional, one-dimensional, two-dimensional and three-dimensional nanostructured materials for advanced electrochemical energy devices. Prog Mater Sci 57:724–803
12. Wang X, Ahmad M, Sun H (2017) Three-dimensional ZnO hierarchical nanostructures: solution phase synthesis and applications. Materials 10:1304–1326
13. Sidorenko AS (2017) Physics, chemistry and biology of functional nanostructures III. Beilstein J Nanotechnol l8:590–591
14. Li S (2008) Fabrication of nanostructured materials for energy applications. Dissertation, Royal Institute of Technology
15. Sivakumar PM, Kodolov VI, Zaikov GE et al (2014) Nanostructure, nanosystems, and nanostructured materials theory, production, and development. CRC Press Taylor & Francis Group, Canada
16. Scholl E (1998) Theory of transport properties of semiconductor nanostructures. Springer Science
17. Garg P, Ghatmale P, Tarwadi K et al (2017) Influence of nanotechnology and the role of nanostructures in biomimetic studies and their potential applications. Biomimetics 2:1–25
18. Khan R, Javed S, Islam M (2018) Hierarchical nanostructures of titanium dioxide: synthesis and applications. IntechOpen
19. Mali SS, Betty CA, Bhosale PN (2014) From nanocorals to nanorods to nanoflowers nanoarchitecture for efficient dye-sensitized solar cells at relatively low film thickness: all hydrothermal process. Sci Rep 4:5451–5458

20. Mohammadinejad R, Karimi S, Iravani S et al (2016) Plant-derived nanostructures: types and applications. Green Chem 18:20–52
21. Watson GS, Watson JA, Cribb BW (2017) Diversity of cuticular micro- and nanostructures on insects: properties, functions, and potential applications. Annu Rev Entomol 62:185–205
22. de Almeida Pachioni-Vasconcelos J, Lopes AM, Apolinario AC et al (2016) Nanostructures for protein drug delivery. Biomat Sci 4:205–218
23. Linko V, Ora A, Kostiainen MA (2015) DNA nanostructures as smart drug-delivery vehicles and molecular devices. Trends Biotechnol 586–594
24. Sun HC, Luo Q, Hou C et al (2017) Nanostructures based on protein self-assembly: from hierarchical construction to bioinspired materials. Nanotoday 14:16–41
25. Jang D, Meza LR, Greer F et al (2013) Fabrication and deformation of three-dimensional hollow ceramic nanostructures. Nat Mater 12:893–898
26. Elsabahy M, Heo GS, Lim SM (2015) Polymeric Nanostructures for Imaging and Therapy. Chem Rev 115:10967–11011
27. Dzenis Y (2008) Structural nanocomposites. Science 319:419–420
28. Lazar MA, Tadvani JK, Tung WS (2010) Nanostructured thin films as functional coatings. In: IOP conference series: materials science and engineering, vol 12, pp 012017–012024
29. Jie Z, Li DX, Hosamani SM (2017) Computation of K-Indices for certain nanostructures. J Comput Theor Nanosci 14:1784–1787

Chapter 4
Physics and Chemistry of Nanostructures

Abstract Nanostructural materials are often in a metastable state. Their detailed atomic configuration depends on nanomaterial fabrication. Therefore, the properties of nanostructures can be widely adjustable by changing their size, shape and processing conditions. The below chapter discuss briefly about physics and chemistry of nanostructured materials.

4.1 Nanostructured Materials

Nanostructures are unique materials whose size of primary structure has been engineered at the nanometer scale. By the demands of miniaturization, increasing efforts are being made to synthesize, understand and apply the materials with reduced dimensions. The interest in nanomaterials is growing at a dramatic rate due to realization that reduced dimensions in nanometer regime can alter and improve the properties of materials. Nanomaterials have attracted widespread attention because of their specific features that differ from bulk materials. This has opened up with tailor-made materials enormous possibilities to meet challenges in different disciplines like medicine, biotechnology, optoelectronics, engineering etc. of new, effective and efficient devices, drugs or tools. It is expected that the emerging technology will focus on challenges of miniaturization and energy saving, covering different technologies that are existing today [1].

Nanostructured materials are a new class of materials which provide one of the greatest potentials for improving performance and extended capabilities of products in a number of industrial sectors, including the aerospace, tooling, automotive, recording, cosmetics, electric motor, duplication, and refrigeration industries. Encompassed by this class of materials are multilayers, nanocrystalline materials and nanocomposites [2]. Their uniqueness is due partially to the very large percentage of atoms at interfaces and partially to quantum confinement effects. For bulk materials, the intrinsic physical properties, such as density, conductivity and chemical reactivity, are independent of their sizes. For example, if a one-meter Cu wire is cut into a few pieces; those intrinsic properties of the shorter wires remain the same as in the original wire. If the dividing process is repeated again and again, this invariance

T. D. Thangadurai et al., *Nanostructured Materials*, Engineering Materials,
https://doi.org/10.1007/978-3-030-26145-0_4

cannot be kept indefinitely. Certainly, the properties are changed greatly when the wire is divided into individual Cu atoms [3]. Significant property changes often start when we get down to the nanoscales. The following phenomena critically affect the properties of nanostructural materials:

Quantum confinement: the confinement of electrons in the nanoscale dimensions result in quantization of energy and momentum, and reduced dimensionality of electronic states.

Quantum coherence: certain phase relation of wave function is preserved for electrons moving in a nanostructure, so wave interference effect must be considered. But in nanostructures, generally the quantum coherence is not maintained perfectly as in atoms and molecules. The coherence is often disrupted to some extent by defects in the nanostructures. Therefore, both quantum coherent and de-coherent effects have to be considered, which often makes the description of electronic motion in a nanostructure more complicated than in the extreme cases.

Surface/interface effects: a significant fraction (even the majority) of atoms in nanostructure is located at and near the surfaces or interfaces. The mechanic, thermodynamic, electronic, magnetic, optical and chemical states of these atoms can be quite different than those interior atoms [4]. These factors play roles to various degrees (but not 100%) of importance. For example, the confinement and the coherent effects are not as complete as that in an atom. Both the crystalline (bulk) states and the surface/interface states cannot be ignored in nanoscale structures. The different mixture of atomic/molecular, mesoscopic and macroscopic characters make the properties of nanostructures vary dramatically. Nanostructural materials are often in a metastable state. Their detailed atomic configuration depends sensitively on the kinetic processes in which they are fabricated. Therefore, the properties of nanostructures can be widely adjustable by changing their size, shape and processing conditions. The situation is similar to molecular behavior in chemistry (e.g., N vs. N_2) in certain aspect. Because of the rich and often surprising outcomes, it will be extremely interesting and challenging to play with nanostructural systems [5].

Only a proper understanding of the dependence of a given property on the particle size can lead to design of the nanostructured material for the related application. It is also important to understand when a material could be considered as nanostructured. Although one can in principle classify materials with grain size less than 100 nm as nanostructured, several properties such as optical and vibrational properties do not differ much from the corresponding bulk value unless the grain/particle size is less than typically 20 nm [6]. In view of this it is reasonable to treat a material with a grain size smaller than a certain value as nanostructured only if the property of interest differs from the bulk value at least by a few percent. It is also possible that a material with nanometer grain size may behave as nanostructured for a specific property while it could act like bulk for other properties. In addition to the grain size, the properties of the nanostructured materials may sometimes depend on the method of their synthesis. Generally nanostructured materials are synthesized in one of the three forms: (a) as isolated or loosely connected nanoparticles in the form of powder, (b) as composites of nanoparticles dispersed in another host, or (c) as compact collection of nanograins as pellets or thin films [7].

The physical and chemical properties of nanostructures are distinctly different from those of a single atom and bulk matter with the same chemical composition. These differences between nanomaterials and the molecular and condensed-phase materials pertain to the spatial structures and shapes, phase changes, energetics, electronic structure, chemical reactivity, and catalytic properties of large, finite systems, and their assemblies. Some of the important issues in nanoscience relate to size effects, shape phenomena, quantum confinement, and response to external electric and optical excitations of individual and coupled finite systems. Size effects are an essential aspect of nanomaterials. The effects determined by size pertain to the evolution of structural, thermodynamic, electronic, spectroscopic, and chemical features of these finite systems with increasing size. Size effects are of two types: one is concerned with specific size effects and the other with size-scaling applicable to relatively larger nanostructures [8] (Fig. 4.1).

For nanostructured materials, like nanomaterials, morphology has special significance since form, in this case, dictates physical and chemical properties. Unlike bulk materials, properties of nanomaterials are strongly correlated to shape. This shape is attained during growth through a self-assembling process dictated by the interplay of size and molecular interactions. Deviations from bulk properties become prominent as the size of nanomaterials starts to be comparable to the size of constituent molecules or to some other characteristic length scale like electron mean-free path. In a typical application, one deals with a collection of nanomaterials, which may be dispersed in a matrix forming a composite material. Properties of this nanocomposite are controlled not only by morphology of individual nanomaterials, but also by the nature of interactions, which, in turn, is determined by the distribution of the nanomaterials in the matrix [9]. Thin films and multilayer structures may introduce a 1D confinement effect in addition to the confinement effect arising due to size of

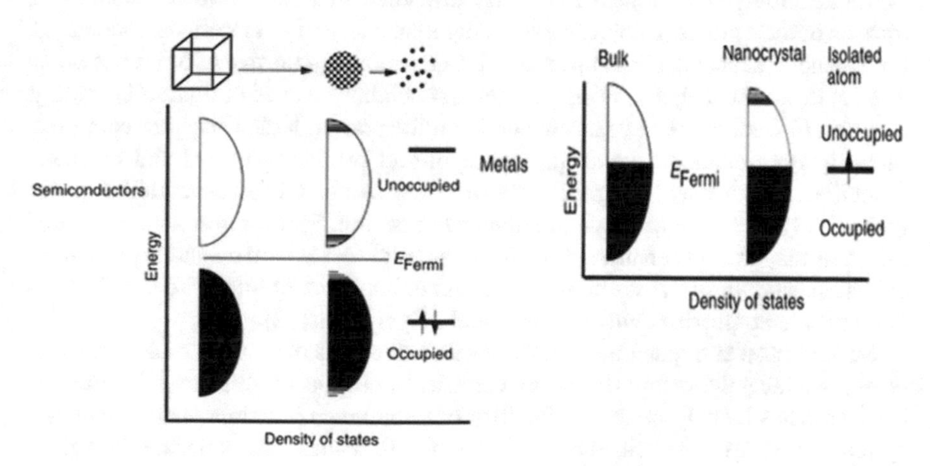

Fig. 4.1 Density of states for metal and semiconductor nanocrystals compared to those of the bulk and of isolated atoms. Reprinted with permission from © Basics of Nanocrystals, Springer, Berlin, Heidelberg (2007)

individual nanomaterials in these films. For example, in a thin film, the presence of different nanomaterials such as nanodots, nanowires, and nanosheets can introduce an additional 3D, 2D, and 1D confinement effect, respectively. In a multilayer structure, apart from atomic arrangements of deposited materials, the presence of interfaces defines an additional 1D periodicity.

The additional 1D potential and associated interfaces of a multilayer modify all physical properties of the deposited materials considerably, and the properties can be tuned by controlling the shape of the 1D potential. In semiconductors, the directional nature of chemical bonds causes formation of extremely high quality epitaxial multilayers. In metals, however, bonding is less directional, and it is more difficult to achieve comparable perfection. Nevertheless, advanced crystal growth methods such as molecular beam epitaxy (MBE) and low-pressure metal organic vapor phase epitaxy (MOVPE) have now made possible the sequential monolayer-by-monolayer deposition not only of semiconductors but also for metals and insulators, and practically any combinations thereof [10]. Multilayered organic and metal-organic films are being studied actively to form model systems for biophysical application and to understand the self-assembling mechanism observed in physics and biology. One of the easiest ways to achieve such a film with good ordering in the direction of growth is the Langmuir–Blodgett (LB) technique. The LB films are very convenient systems for studying melting of 2D solids, which is expected to be a continuous transition as opposed to melting of conventional 3D solids. Nanometer-sized semiconductor particles that exhibit quantum confinement effect in band structure can be formed by exposing suitable LB films to reactive gases [11].

The development of nanostructured materials has led to a new class of materials that are single- or multiphase polycrystals with microstructural features, i.e., particle or grain sizes, layer thicknesses, or domain sizes, in the nanometer range. Owing to the extremely small dimensions, nanostructured materials have an appreciable fraction of their atoms in defect environments such as grain or interface boundaries. For example, nanocrystalline material with an average grain size of 5 nm has about 50% of its atoms within the first two nearest-neighbor planes of a grain boundary, in which distinct atomic displacements from the normal lattice sites are exhibited. These unique features have an important impact on their physical and chemical properties, which may be significantly different compared to conventional coarse grained polycrystals of the same chemical composition. For example, nanostructured materials may exhibit enhanced diffusivity, superior soft or hard magnetic properties, enhanced catalytic activity, ultrahigh strength or hardness, or improved ductility and toughness in comparison with conventional polycrystals [12].

Superlattices and quantum wells were introduced as man-made quantum structures to engineer the quantum states for electrical and optical applications. In analysis, the idea relies heavily on the availability of good heterojunctions, lattice matched systems, and later, the strained layered systems. To realize quantum states in a given geometry, the size must be smaller or comparable to the coherence length of electrons, in order to exhibit quantum interference. This requirement eliminates doping as an effective means to achieve confinement, except at low temperatures, because doping comes from charge separation which results in barriers generally far exceeding the coherence length of electrons at room temperatures. On the other hand, band-edge alignment of a heterojunction provides abrupt barrier height. This short range potential is the consequence of higher orders multiples in the atomic potentials [13].

The advent of reliable production of nanostructures has opened a frontier in material science. As the size of these structures or devices approaches the nano-meter scale, the laws of quantum mechanics come into play. Quantum dot structures are being considered for a variety of technological applications ranging from semiconductor electronics to biological applications including optical devices, quantum communications and quantum computing. The understanding of the quantum structure's electronic properties and quantum confinement is of paramount importance. The basic concept of quantum confinement comes from the interplay of two fundamental principles of quantum mechanics; namely the electronic system must obey the Schrodinger equation and also follow the de Broglie momentum-wavelength relationship. It is desirable to be able to position the quantum dots in space in a determined order or pattern. Nucleation site engineering can be used to achieve the spatial ordering of self-assembled quantum dots. For example, selective area epitaxy and the anisotropic indium mobility have been used to create narrow ridges and promote the formation of quantum dots positioned in a single row on such ridges [14] (Fig. 4.2).

Nanostructured materials of varying composition continue to entice the materials research community with the promise of innumerable practical applications as well as advancing an understanding of their fundamentals. The exploration of size and

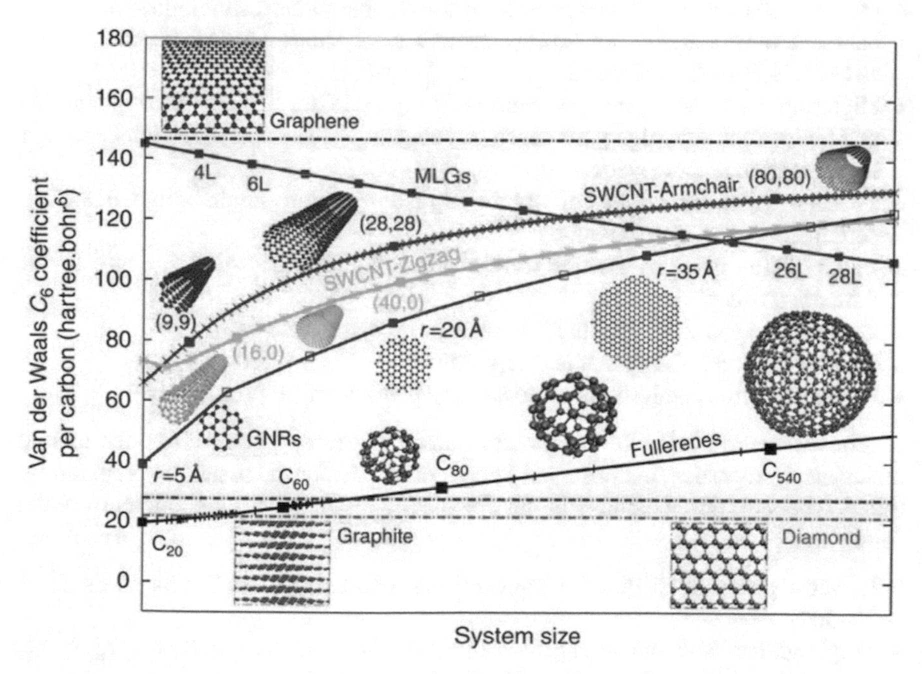

Fig. 4.2 Scaling laws of quantum coefficient. Reprinted with permission from © Nature Communication (2013)

structure influences on materials properties such as dielectric constant, conductivity, and luminescence has developed into a burgeoning new subdiscipline of materials science Synthetic strategies routinely achieve complex subarchitectures, sending pore sizes plummeting into the nanoscale regime, where the ratio of surface atoms to bulk increases to significant proportions. It is in this size regime that the surface properties of these materials become of paramount significance. Thc chemical properties of the interfacial atoms directly influence the stability, reactivity, and overall utility of a material in a given application. Although the interfacial regions have proven to play such a dominant role in the materials properties of these systems, only recently has their chemical reactivity and stability been intricately explored [15].

The nanostructures have higher surface areas than do conventional materials. The impact of nanostructure on the properties of high surface area materials is an area of increasing importance to understanding, creating, and improving materials for diverse applications. High surface areas can be attained either by fabricating small particles or clusters where the surface-to-volume ratio of each particle is high, or by creating materials where the void surface area is high compared to the amount of bulk support material. Materials such are highly dispersed supported metal catalysts and gas phase clusters fall into the former category and microporous (nanometer-pored) materials such as zeolites, high surface area inorganic oxides, porous carbons, and amorphous silica. The use of the nanostructure approach to high surface area materials may have significant impact:

- Microporous materials for energy storage and separations technologies, including nanostructured materials for highly selective adsorption/separation processes such as H_2O, H_2S, or CO_2 removal.
- High Capacity, low volume gas storage of H_2, and CH_4 for fuel cell applications and high selectivity; high permeance gas separations such as O_2 enrichment; and H_2 separation and recovery.
- Thermal barrier materials for use in high temperature engines understanding certain atmospheric reactions.
- Incorporation into construction industry materials for improved strength or for fault diagnostics.
- Battery or capacitor elements for new or improved operation.
- Biochemical and pharmaceutical separations.
- Product-specific catalysts for almost every petrochemical process [16].

The mechanical behavior of nanostructured materials originates from the unique mechanical properties first observed and/or predicted for the materials prepared by the gas condensation method. Among these early observations/predictions were the following:

- Lower elastic moduli than for conventional grain size materials-by as much as 30–50%.
- Very high hardness and strength-hardness values for nanocrystalline pure metals are 2–7 times higher than those of larger grained (>111) metals.

- A negative Hall-Petch slope, i.e., decreasing hardness with decreasing grain size in the nanoscale grain size regime.
- Ductility-perhaps superplastic behavior-at low homologous temperatures in brittle ceramics or intermetallics with nanoscale grain sizes, believed due to diffusional deformation mechanisms [17].

References

1. Harik VM, Salas MD (2003) Trends in nanoscale mechanics. ICASE LaRC interdisciplinary series in science and engineering, 9th edn. Springer
2. Champion Y, Fecht HJ (2004) Nano-architectured and nanostructured materials. WILEY-VCH Verlag GmbH & Co, KGaA, Weinheim
3. Nalwa HS (2000) Handbook of nanostructured materials and nanotechnology. Academic Press
4. Lockwood DJ (2004) Nanostructure science and technology. Kluwer Academic Publishers, Boston
5. Yan H, Yang P (2003) Semiconductor nanowires: functional building blocks for nanotechnology. World Scientific Publishing Co. Pte. Ltd., Singapore
6. Ramrakhiani M (2012) Nanostructures and their applications. Recent Res Sci Technol 4:14–19
7. Arora AK, Rajalakshmi M, Ravindran TR (2004) Phonon confinement in nanostructured materials. Encyclopedia Nanosci Nanotechnol 8:499–512
8. Biener J, Wittstock A, Baumann TF (2009) Surface chemistry in nanoscale materials. Materials 2:2404–2428
9. Hollingsworth MD (2002) Crystal engineering: from structure to function. Science 295:2410–2413
10. Weissbuch C, Vintner B (1991) Quantum semiconductor structures: fundamentals and applications. Academic, Boston
11. Sanyal MK, Datta A, Hazra S (2002) Morphology of nanostructured materials. Pure Appl Chem 74:1553–1570
12. Eckert J (2007) Structure formation and mechanical behavior of two-phase nanostructured materials. In: Nanostructured materials, pp 423–526
13. Tsu R, Zhang Q (2007) Nanostructured electronics and optoelectronic materials. In: Nanostructured materials, pp 527–568
14. Denison AB, Hope-Weeks LJ, Meulenberg RW et al (2004) Quantum dots. In: Introduction to nanoscale science and technology. Kluwer Academic Publishers, New York
15. Porter LA, Buriak JM (2003) Harnessing synthetic versatility toward intelligent interfacial design: organic functionalization of nanostructured silicon surfaces. In: The chemistry of nanostructured materials. World Scientific Publishing Co. Pte. Ltd., Singapore
16. Siegel RW, Hu E, Roco MC (1999) Nanostructure science and technology. WTEC panel on nanoparticles, nanostructured materials, and nanodevices. Springer, Dordrecht
17. Gleiter H (2000) Nanostructured materials: basic concepts and microstructure. Acta Mater 48:1–29

Chapter 5
Quantum Effects, CNTs, Fullerenes and Dendritic Structures

Abstract Nanostructural materials have wide classification of structures based on their fabrication methods. Fullerenes are close-caged molecules containing only hexagonal and pentagonal interatomic bonding networks. Carbon nanotubes are large, linear fullerenes with aspect ratios as large as 10^3–10^5. Nanotubes as many derivatives like nanocones, nanosprings, etc. The chapter detailed about different nanostructures and its properties.

5.1 Fullerenes Structures

Fullerenes a interrelated research field including organic transformations of these all-carbon hollow-cluster materials has emerged. C_{60} has been the most thoroughly studied member of fullerenes because it (1) is produced abundantly in the carbon soot by the arc discharge of graphite electrodes, (2) has high symmetry, (3) is less expensive, (4) is relatively inert under mild conditions, and (5) shows negligible toxicity. Electronically, C_{60} is described as having a closed-shell configuration consisting of 30 bonding molecular orbitals with 60p electrons, which give rise to a completely full fivefold degenerated highest occupied molecular orbital that is energetically located approximately 1.5–2.0 eV lower than the corresponding antibonding lowest unoccupied molecular orbital (LUMO) one [1]. The first electron in the reduction of C_{60} is added to a triply degenerate t_{1u} unoccupied molecular orbital and is highly delocalized. This threefold-degeneracy, together with the low-energy possession of the LUMO, makes C_{60} a fairly good electron acceptor with the ability of reversibly gaining up to six electrons upon reduction. The facile reduction contrasts with its difficult oxidation. Only the first three reversible oxidation waves have been observed. This high degree of symmetry in the arrangement of the molecular orbitals of C_{60} provides the foundation for a plethora of intriguing physicochemical, electronic, and magnetic properties. Semiconducting, magnetic, and superconducting properties of unmodified C_{60} have been intensively investigated; however, these properties remain

T. D. Thangadurai et al., *Nanostructured Materials*, Engineering Materials,
https://doi.org/10.1007/978-3-030-26145-0_5

to be explored for functionalized fullerenes. On the other hand, nonlinear optical and photophysical properties of functionalized fullerene materials have been already under investigation. The skeleton of C_{60} consists of 20 hexagonal and 12 pentagonal rings fused all together.

The chemical reactivity of C_{60} is that of a strained electron deficient polyalkene, with rather localized double bonds. Cyclo addition reactions have been widely applied for the functionalization of fullerenes. Usually, addition occurs across ring junctions where electron density is much higher than at the ones. Other reactions that take place on the spherically shaped carbon core of fullerenes involve additions of nucleophiles, free radicals, and carbenes as well as η^2-complexation with various transition metal elements. In such fullerene adducts, the functionalized carbon atoms change their hybridization from a trigonal sp^2 to a less strained tetrahedral sp^3 configuration; thus, the primary driving force for addition reactions is the relief of strain in the fullerene cage. Furthermore, the regiochemistry of the addition is governed by the minimization of the 5,6-double bonds within the fullerene skeleton [2]. Therefore, any 1,2-addition reactions occur to produce ringclosed 6,6-adducts, having two sp^3 carbon atoms on the fullerene framework. However, sometimes ring-opened 6,5-adducts (fulleroids) are formed, keeping all fullerene carbon atoms sp^2-hybridized. The main advantage gained upon functionalization of fullerenes is a substantial increase in their solubility. The existence of a great diversity of synthetic protocols combined with the high number of chemical reactions that have been mainly applied to C_{60} has led to the formation of a wide variety of functionalized fullerenes. The special characteristics of the added groups, coupled with the unique structural, physicochemical, and electronic properties of fullerenes (which in most of the cases are retained after functionalization), have aided the development of new materials with tremendous potential in fascinating and widespread technological applications such as electronic and optoelectronic devices, light-emitting diodes, photovoltaics, and thermotropic liquid crystals [3].

New supramolecular motifs of fullerenes with concave–convex structures have been developed very recently by π-stack interactions. It comprises interactions between fullerenes and different curved carbon structures, such as nanorings or nanotubes. Both structures have π-orbitals oriented radially, as well as the fullerenes. Nanorings are closed loops formed by conjugated π-orbitals. They are very interesting from the point of view that they can host different molecules depending on the size of the ring. Fullerenes can be hosted by nanorings. They prepared [6]- and [8]paraphenyleneacetylene and studied their complexation with C_{60} and with bis(ethoxycarbonyl)methanofullerene.

The experiments with C_{60} showed the formation of a 1:1 supramolecular complex with [6]paraphenyleneacetylene (Fig. 5.1) and the association constant was determined ($K_a = 16{,}000\ M^{-1}$). The same measurements were attempted with bis(ethoxycarbonyl)methanofullerene but the changes in absorption were too small to measure a reliable association constant. On the other hand, it was possible to obtain an X-ray structure from the fullerene monoadduct and [6]paraphenyleneacetylene, confirming the structure of the complex. The experiments with [8]paraphenyleneacetylene showed none or very small complexation with C_{60}. In a different article from

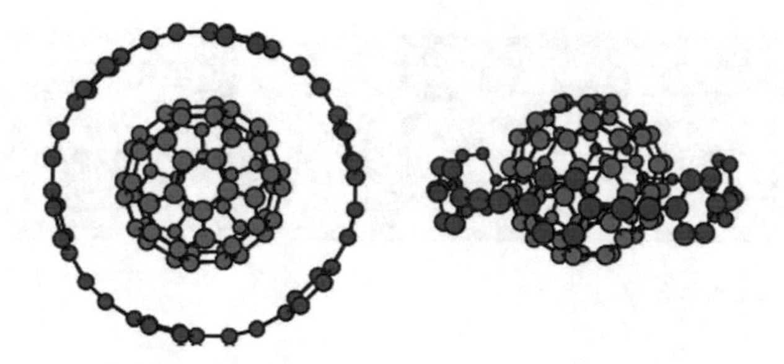

Fig. 5.1 C_{60}-nanoring complex

the same authors, the complexation of C_{60} and C_{70} with [6]- and [7]paraphenyleneacetylene was studied. Also other complexation studies were carried out with the novel isomers of [7]paraphenyleneacetylene that have 1,4- or 2,6-naphtylene units diametrically opposed [4]. The calculation of the association constants was attempted but they were too large to be determined precisely (>5 × 10^4), which indicated that very stable complexes were formed. Fluorescent measurements were carried out from these complexes, providing a lot of information about the estimated relative stability of the whole series of rings with C_{60} and C_{70}. The results were consistent with the values of the association constants and also showed that the new naphtylene nanorings are very efficient fluorescent sensors for fullerenes. During the synthesis and purification of the different paraphenyleneacetylene and 1,4-naphtyleneacetylene nanorings, it was found that big rings could form inclusion complexes with smaller rings. [6]Paraphenyleneacetylene could be inserted within the larger [9]paraphenyleneacetylene, the association constant could only be determinates at −60 °C ($K_a = 340 \pm 45\ M^{-1}$). The same behavior was observed with the analogous 1,4-naphtyleneacetylene nanorings showing much higher association constants at −60 °C ($K_a = 11{,}000 \pm 1400\ M^{-1}$) because of the larger contact area. In order to prepare complexes including fullerenes, C_{60} was first included within the smaller ring as done in previous experiments and subsequently these complexes were then assembled with the larger ring. NMR experiments proved the formation of these onion-type complexes (Fig. 5.2). The association constants could only be calculated for the [9]paraphenyleneacetylene ([6]paraphenyleneacetylene C_{60}) complex at −

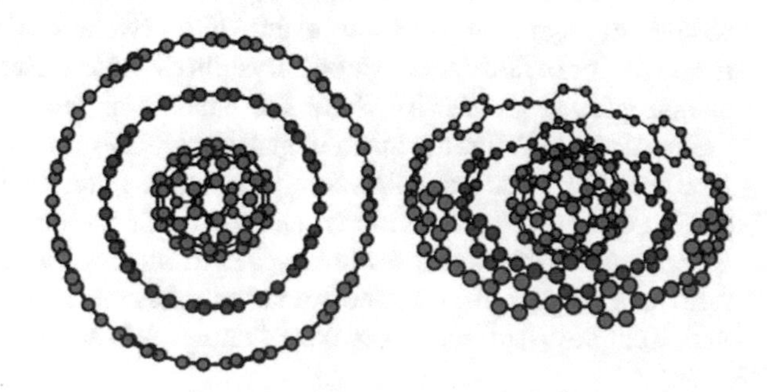

Fig. 5.2 C_{60} Onion type complex

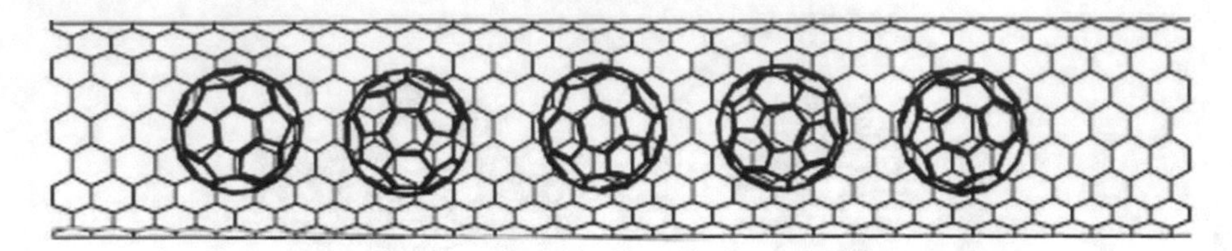

Fig. 5.3 Carbon peapods

60 °C ($K_a = 410 \pm 80\ M^{-1}$). Other interesting structures known as carbon peapods have been developed in the last years. Carbon peapods are single-walled carbon nanotubes (SWCNTs) that are filled with fullerenes (Fig. 5.3). They are considered as a new supramolecular carbon allotrope.

Carbon peapods were first reported in 1998, when they were detected by high-resolution transmission electron microscopy (HRTEM) as side products during the production of carbon nanotubes. Since then, several methods for the efficient filling of empty SWCNTs have been developed. Carbon peapods were prepared from raw nanotubes produced by pulsed laser vaporization. Several defects were introduced on the surface of the nanotubes by refluxing the material in nitric acid, thus facilitating the insertion of fullerenes. Subsequently they were annealed at high temperatures and low pressures. C_{60} was only inserted in 1.3–1.4 nm diameter SWCNTs that are the only ones that permit the preferred graphitic 0.3 nm van der Waals separation between the fullerene and the nanotube [5].

5.2 Nanostructures

Common to the synthesis or deposition of bulk and nanocrystalline solids—whether from the liquid or vapor phase—are the concepts of nucleation and growth. Nucleation of the solid phase on a surface or in solution, by definition involves the initial formation of clusters of atoms due to supersaturation and subsequent precipitation of a selected elemental species or compound. Viewing synthesis of nanostructures as the early stage of bulk crystal growth, it should be emphasized that in contrast to 0-D and 2-D growth, the formation of 1-D structures is far less likely to occur in the absence of a symmetry-breaking event, of interfacial strain between the substrate and the nuclei, or of differences among crystallographic surfaces that can lead to strongly anisotropic growth rates [6]. Some 1-D nanostructured materials may be grown using methods in which there is inherent growth rate anisotropy to the crystalline structure. However, for other technologically important materials (e.g., Si, GaAs) crystalline structure alone often does not typically provide the necessary driving force to overcome more energetically favorable growth modes. Thus, a central challenge in the rational synthesis of 1-D nanostructured materials is to employ growth conditions, which employ a symmetry-breaking feature. Additionally, for some applications and fundamental studies, methods that permit dislocation and twin-free structures or precise control of diameter and length, and size dispersion may be important. In some situations, methods that produce periodic or non-periodic arrays of aligned nanostructures may be desired. Finally, the need to adapt a process for other material systems or compositions to carry out growths at lower temperatures, to have compatibility with existing processes, and adaptability for multi-component architectures, are important considerations. The methods of synthesizing single-crystalline 1-D nanostructured materials from the bottom-up may be classified into those that employ inherent crystalline anisotropy, those that exploit a metal nanoclusters catalyst to facilitate 1-D growth, those that rely on templates, and those that rely on kinetic control and capping agents. Specifically, kinetic control is gained by monomer concentration or by multiple surfactants, bringing about conditions that strongly favor anisotropic growth. Though not discussed here, self- or directed-assembly of 0-D nanostructures into 1-D nanostructures, and size reduction of 1-D structures are additional process routes. Alternatively, one can classify the strategies into chemical solution-based syntheses, and those via vapor condensation routes, by metal organic-vapor phase, by thermal means, or by laser ablation [7].

5.3 Laser-Assisted Metal-Catalyzed Nanowire Growth

In laser-assisted metal-catalyzed nanowire growth (LCG), laser ablation of a target containing thc desired NW material(s) provides the source; the process is based, in part, on pulsed laser deposition (PLD), a common route for producing a wide variety of different classes of inorganic thin-film materials. The LCG method involves the flowing of an inert or carrier gas through a furnace, and the use of a laser-ablation target and of a growth substrate with metal nanocluster-catalyst particles. Alternatively, the catalyst material may be contained within the target, and may be prepared by sintering appropriate components. In general, laser wavelengths λ_e for LCG are chosen so that the optical absorption depth is shallow in order to maximize the effect of heating over a relatively small volume. Often, for growth of semiconductor NWs, an Nd:YAG ($\lambda_e = 1064$ nm, or frequency doubled or tripled) or an excimer laser (e.g., ArF, KrF, or XeF) is used. Ablation targets are typically placed at the upstream end, and in cases where heating of the target is needed, within the hot zone. Alternatively, if the required selected NW growth and furnace temperature is higher than the melting point of the target, targets are placed just outside the furnace. The LCG method is a highly versatile route for obtaining a wide range of NW materials—many materials can be prepared via laser ablation. It is also attractive in that the laser ablation route eliminates the need for gaseous precursors, many of which are toxic and, in some cases, also pyrophoric. Many of the target materials are widely available or can be prepared, for example, by sintering of particle components. Moreover, the method can be used in conjunction with a vapor deposition process to yield NWs having solid solutions and periodic segments of different composition. The LCG methods can be pursued as a synthesis route even in the absence of detailed phase diagrams. Despite its versatility, a shortcoming of the LCG method for the growth of semiconducting NWs (when used alone) is that the targets themselves typically must contain all of the components required for the NW. This limits the ability to control the introduction of dopants or other dilute impurities during growth, or the ability to adjust the composition. Moreover, to date, precise control of diameter, diameter dispersion, and synthesis of ultra-narrow (below ~10 nm) NWs has not been demonstrated with LCG [8].

5.4 Hierarchal Complexity in 1-D Nanostructures

The important features of inorganic-nanomaterial synthesis by surfactant-mediated chemical routes are the possibility of controlling the size and shape of nanostructures, thereby effectively manipulating their properties. The growth range of nanostructures by varying the relative concentrations of two types of surfactants–TOPO and hexylphosphonic acid (HPA) and systematically varying the monomer concentration with time. The HPA serves to accentuate the difference between growth rates on different crystal faces. Different systematic variations yield long rods, arrow-shaped NCs, teardrop-shaped NCs, tetrapods, and dendritic tetrapods. The differences between wurtzite and rocksalt free energies and surface energetics govern the evolution of some of these structures-particularly stacking faults in the tetrapods. This explains the topology in terms of abrupt changes in the crystal structure, from wurtzite in the rod segments to rock salt in the sphere-like vertices.

The investigations indicated that multiphase description of the tetrapod, and the corresponding model of their formation may not be valid: rather, it seems that the tetrapod nanostructures may consist of single phase, and that it is likely that the formation of twins plays a significant role in the unique evolution of their topologies. The control of the relative axial and radial growth rates is the basis for the synthesis of coaxial or core–shell nanostructures as discussed above. In our laboratory at Drexel, we have demonstrated the growth of apex-angle-controlled crystalline Si nanocones (Si NCs) and Ge nanocones (Ge NCs) of diamond-hexagonal phase through simultaneous control of axial and radial growth rates using metal catalyzed CVD; we have also synthesized single-crystalline Ge nanosprings and characterized its composition and crystallinity with electron microscopy and Raman scattering. As shown in Fig. 5.4, the Si NCs are tapered polyhedral, possessing hexagonal cross-sections. They are typically several microns in base diameter, and the radius of curvature at the tip is as small as 1–2 nm. Significantly, these nanocones are of the diamond hexagonal polymorph, and the taper angles can be tuned during growth. These Si NCs may offer new opportunities as scanning probes and as central components in single-molecule sensing [9] (Fig. 5.5).

The integration of two or more semiconductor or other materials and the controlled incorporation of dopants within individual components of nanostructures are essential characteristics of the advancing development of nanowires (NWs) as building blocks for nanotechnology. Homo- and heterojunctions within individual NWs

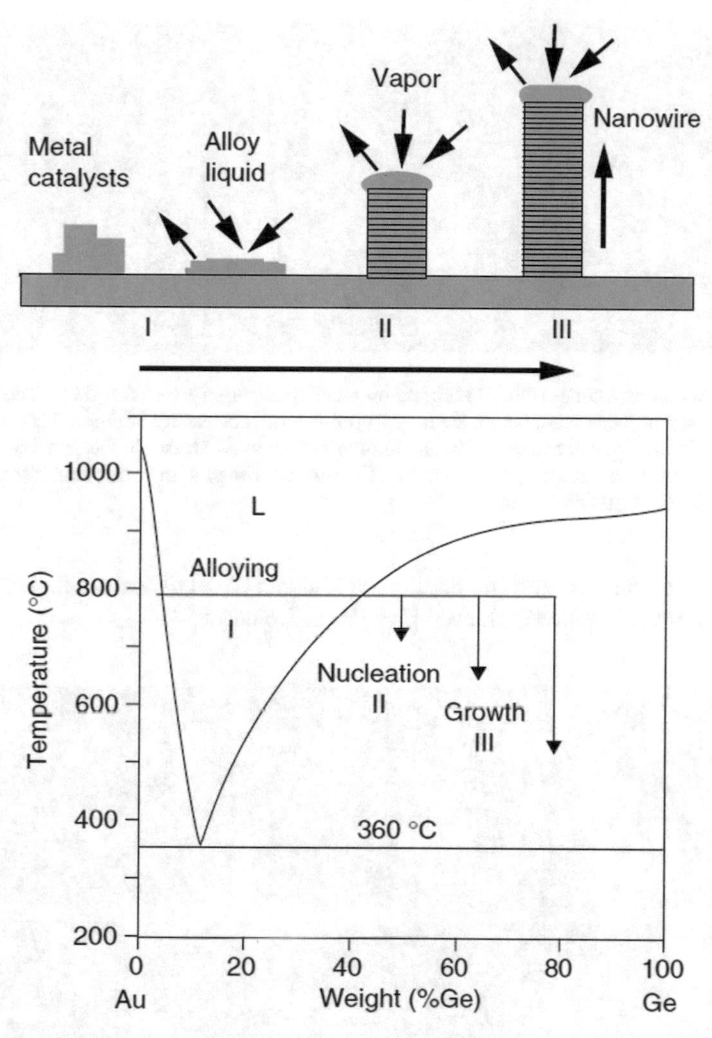

Fig. 5.4 Different phases of NW growth. Reprinted with permission from © Top Curr Chem, 226, 119–172 (2003)

enable fabrication of bipolar devices within individual NWs, opening up possibilities for light-emitting diodes, photodetectors, and band gap-engineered devices within individual NWs, and even NW-based superlattices. Some synthesis methods permit formation of hierarchal nanostructures, including NWs with segments composed of different materials. For most electronic or photonic device applications, synthesis of single-crystalline components with some degree of control of composition at interfaces is desired. The synthesis of hierarchal nanostructures via metal-catalyzed and catalyst-free VLS and solid–liquid–solid methods has been demonstrated successfully in several semiconducting and functional-oxide material systems. Specifically,

Fig. 5.5 Silicon nanocones (Si NCs) formed by metal nanocluster-catalyzed CVD. The left panel shows an array of nanocones, which are tapered polyhedral, possessing hexagonal cross-sections. These Si NCs are typically several microns in base diameter. As shown in the panel on the right, the radius of curvature at the tip is as small as 1–2 nm. Reprinted with permission from © Taylor & Francis Group, CRC press (2006)

modulation of composition and doping in axial and radial directions in a number of material systems has been achieved [10] (Figs. 5.6 and 5.7).

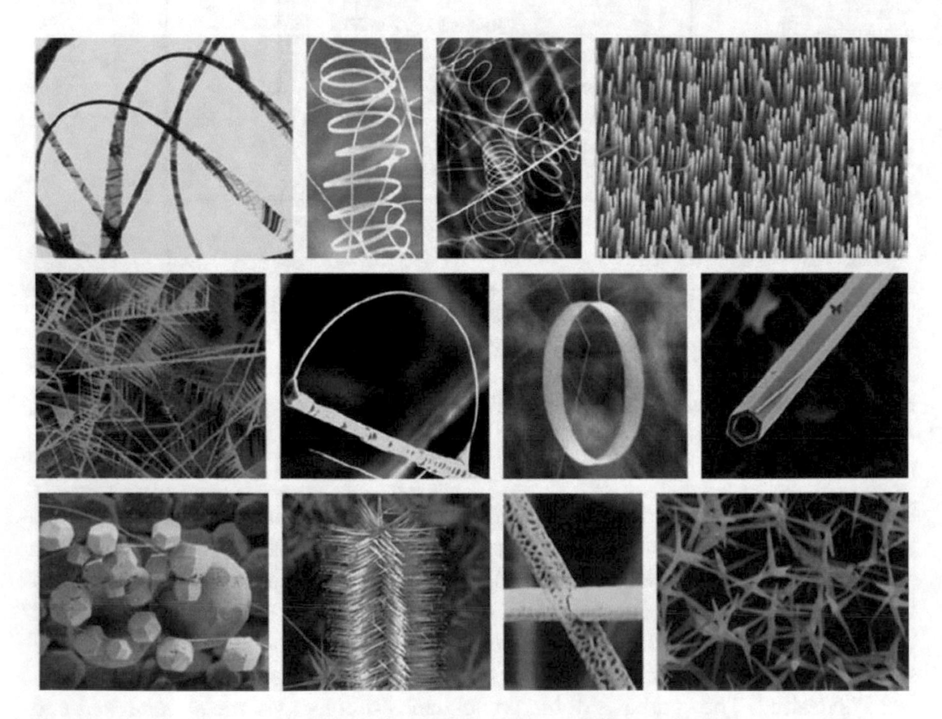

Fig. 5.6 Representative polar surface-dominated, single-crystalline, zinc oxide 1-D nanostructures. Reprinted with permission from © Thin Solid Films, 562, 291–298 (2014)

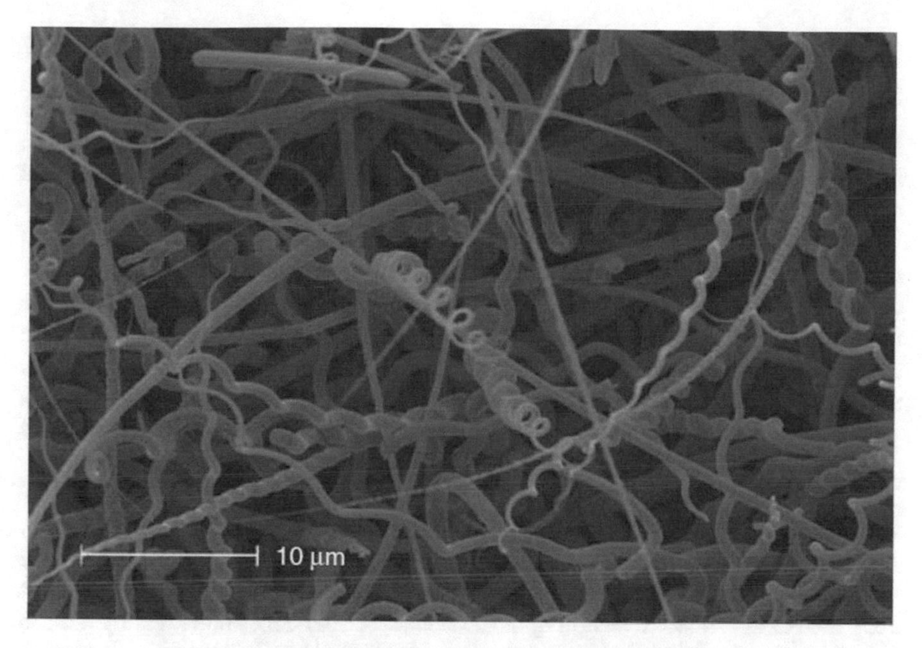

Fig. 5.7 Scanning electron micrograph of crystalline germanium nanosprings. The formation of nanosprings having inversion symmetry may generate further debate on the formation mechanisms of these and similar structures, in both polar and nonpolar crystalline materials. Reprinted with permission from © Applied Physics Letters, 88, 193105–193107 (2006)

5.5 Mechanical and Thermal Properties

The mechanical properties of materials are not scale-invariant. For example, the materials that undergo plastic deformation, the so-called Hall and Petch behavior—the increase in the hardness and yield strength of a polycrystalline material with reduction in grain size—involves the relative increase in the area of grain boundaries and the associated piling up of dislocations. The reduction in grain size leads to more impediments to dislocation motion and results in a scale-dependent toughness, though there exists a finite nanoscale size for which the toughness reaches a maximum. With respect to ceramic nanomaterials, relatively little has been reported on the scaling of mechanical properties in the semiconducting and oxide NW or nanobelt materials [11].

Thermal conduction via phonons is expected to be significantly reduced in narrow NWs. When the dimensions of a NW approach the mean free path of phonons, the thermal conductivity is reduced relative to bulk values by scattering of phonons by surfaces and interfaces. The miniaturization of electronic and photonic devices, e.g., the higher number density of transistors, is accompanied by new challenges in effective thermal management. A figure of merit for thermoelectric materials, however, involves the unusual combination of poor thermal conductivity with high electrical conductivity. Yang and his group at Berkeley have demonstrated that axial periodicity in Si/SiGe axial superlattice NWs may be highly effective in tuning NW thermal conductivities for thermoelectric applications; they reported the reduction in the thermal conductivity of Si/SiGe superlattice NWs by five times, compared with elemental undoped Si NWs of the same diameter. Raman scattering is one of the most powerful experimental tools for characterizing lattice dynamics and corresponding thermodynamic properties in solids. A number of investigations have been performed to characterize the effects of finite size and wire-like geometry on the confinement of phonons. Theoretical predictions indicate that for Si NWs of diameters <~20 nm, phonon dispersions become altered and group velocities are lowered relative to bulk values. Reduction in NW diameter is accompanied by a downshifting and asymmetric broadening of the Raman line shape, and has been attributed to the confinement of optical phonon(s). In perfect bulk single crystals with no loss of translational symmetry, first-order Raman scattering is restricted to $q = 0$ phonons in accordance with momentum selection rules.

The reduction of size, leads to the relaxation of these selection rules and necessitates an inclusion of a larger fraction of the dispersion curves away from the Γ point. A model for including and averaging $q \rightarrow 0$ phonons was developed and applied by Richter et al. and by Herman and coworkers. This Gaussian-correlation model has been used to describe the sampling of the phonon dispersion, and is widely used to model phonon confinement in nanostructures, including NWs. Enhancements of the model is included to incorporate known-size dispersion and the effects of strain via a Gruneisen parameter. In addition, the effects of lattice strain at and near the NW surface defects, and coupling of free carriers to longitudinal optical phonons in degenerately doped semiconductors may also affect the Raman line shapes and zone-center phonon energies. The use of hierarchal nanostructures becomes more prevalent in nanoelectronics and nanophotonics, Raman scattering will continue to play a unique role in the evaluation of 1-D nanostructure materials and devices, including, for example, crystal structure and quality, interfacial strain, thermal management, and strongly correlated electron behavior [12].

5.6 Electronic Properties of Nanowires

Since the invention of modulation doping and the higher electron mobility transistors by, precise control of the composition in semiconductors remains a critical component to 2-D electronic and photonic devices. In a seminal 1980 publication, pointed out that the restriction of electronic carriers to 1-D from 2-D or 3-D would result in significantly reduced carrier-scattering rates, owing to the reduction in the possible k-space points accessible to carriers. In general, electronic-carrier mobility in real systems can be affected by the scattering of carriers in a number of ways: scattering by other carriers, by surfaces, by interfacial roughness, by acoustic phonons, optical phonons, impurities, and by plasmons. Significant theoretical and experimental work involving electronic transport in CNTs has helped in distinguishing ballistic and diffusive modes of transport. However, for free-standing semiconductor NWs, experimental and theoretical consensus of carrier-scattering mechanisms that are most significant in single- and multicomponent coaxial semiconductor NWs is less clear. In some cases, carrier mobilities in Si NW-field effect transistors and transconductance values have been reported to exceed those associated with conventional Si planar technology devices. Even if the mobility of carriers in semiconductor NWs in a given device is lower than that in the corresponding bulk material, there are a number of other possible advantages to 1-D transistor devices over their 2-D counterparts. With respect to electronic properties, NWs represent an important link between bulk and molecular materials. The electronic properties of many bulk semiconductors, including oxide semiconductors, are well known, and the electronic-band structures have, in many cases, been modeled in great detail [13]. Systematic control of NW diameter enables systematic investigation of the effects of dimensionality on electronic transport. In general, the diameter below which electronic transport is significantly altered is related to the degree of confinement of carriers and excitons, the Fermi wavelength and Coulomb interactions, although NWs with larger diameters still possess significant surface-to-volume ratios that can affect electronic transport via surface scattering. Semiconductor NWs, like CNTs, are currently being investigated for elements in nanoscale spintronic devices by introducing dilute concentrations of magnetic dopants. Several groups have reported successful doping of semiconductor NWs with manganese, using different methods. There are numerous reports on the field emission properties of transition metal oxide nanowires and semiconductor nanocones. There have been a number of simulations reported of the electronic transport and properties of semiconductor NWs. The strategies can be classified into effective mass and $k \cdot p$ methods, tight-binding theory, and pseudo potential approaches. In the past, the sizes of unit cells required to describe NWs, using first-principles density functional theoretical (DFT) methods, were too computationally expensive to pursue; this has changed of late. DFT study of the structural and electronic properties of [111]-oriented Si-Ge core–shell interface in very small-diameter (1–2 nm) NWs. Significantly, the authors found positive and negative deviations from Vegard's law for compressively strained Ge and tensile-strained Si cores respectively. They also found that the direct-to-indirect transition for the fundamental band gap is found in Ge-core/Si-shell NWs, while Si-core/Ge-shell NWs preserve a direct energy gap almost over the whole compositional range. The earlier study reported the density-functional theoretical simulations of electronic-band structure and optical properties of Si NWs, predicting that NWs having a principal axis along [110] possess a direct gap at the Γ point for wires modeled up to 4.2 nm in diameter. These findings hold significant promise for the potential application of such Group IV single component and coaxial NW heterostructures for a broader range of photonic applications. The formation of addressable 0-D semiconductor structures is important for fundamental studies of electronic transport phenomenon, and for future applications in single-electron transistors, quantum computing, and metrology.

Axial modulation of the composition of NWs may be used to produce precisely located 0-D and 1-D segments within NWs. The authors reported the fabrication of single-electron transistors and resonant-tunneling diodes by producing heterostructural interfaces separated by as few as several monolayers or as many as large as hundreds of nanometers. In addition, the authors have recently used these so-called 1-D/0-D/1-D structures to fabricate nanowire-based, single-electron memory devices [14]. Superlattices in NWs based on these structures may enable extension of the quantum cascade laser as pioneered. Addressability of individual NWs for electronic and optical devices is an important consideration for practical applications: among the developments in this area, reported on the use of NW-crossbar arrays as address decoders. Electronic transport in semiconductor-NW devices has been shown to be a highly sensitive means of selective binding and detection of molecules on NW surfaces functionalized with receptors. Nanowire-based detection of a number of analytes, including the prostate-specific antigen (pSA), label-free, single virus detection has been demonstrated. NW arrays with selectively functionalized NW surfaces represent one of the frontiers in high-sensitivity detection of a broad range of chemical and biological molecules and other species. An overview of this work has been presented in a recent review. Among the most significant recent achievements pertaining to electronic transport are reports of coherent transport in molecular scale semiconductor NWs and the report of formation of a 1-D hole gas in a Ge/Si NW heterostructure. The authors reported room-temperature ballistic hole transport in semiconductor NWs with electrically transparent contacts, opening important further possibilities for new fundamental studies in low-dimensional, strongly correlated carrier gases and high-speed, high-performance nanoelectronic devices based on semiconductor NWs. The proximal superconductivity was reported in semiconductor NWs, in which the electronic transport characteristics of InAs NWs with closely spaced aluminum electrodes were measured at temperatures below the superconducting transition temperature of Al. The NW-electrode interfaces act as mesoscopic Josephson junctions with electrically tunable coupling [15].

5.7 Optical Properties of Nanowires

In NWs, like their 0-D nanostructure and 2-D ultrathin film counterparts, confinement of electronic carriers leads to altered electronic-band structures that are manifested as changes in band-to-band transition emission and absorption energies, and changes in other optical properties according to the degree of confinement. In order to provide a basic description of the effect of the confinement of electronic carriers to within a NW, the electronic wave functions and bound-state energies are obtained by solving the Schrodinger equation under the effective mass model and with the assumption that the NW has a circular cross-section. One of the most exciting developments involving the optical properties of NWs is the demonstration of room-temperature lasing emission in the ultraviolet from individual NWs. The well-faceted ZnO and GaN NWs produced using VLS growth methods and having diameters from 100 to 500 nm support predominantly axial Fabry–Perot waveguide modes, which are separated by $\Delta\lambda = \lambda^2/[2Ln(\lambda)]$, where L is the cavity length and $n(\lambda)$ the group index of refraction. Structures with smaller diameters are prevented from lasing due to diffraction, and photoluminescence emission is lost in the form of surrounding radiation. The lasing in CdS NW optical cavities was also reported. The flexibility of the hierarchal structure of NWs has also been applied to NW lasers. Using the core–shell GaN/$Al_{0.75}Ga_{0.25}N$ configuration, others works provide simultaneous exciton and photon confinement in NW cores having diameters as small as 5 nm much smaller than the normal minimum to avoid the diffraction effects that prevent lasing. The Berkeley group has extended these concepts to a family of NW photonic components, and taken advantage of the ability of these NWs to sustain significant deformation without fracture in order to develop NW-based optical networks and devices. In fact, the large nonlinear optical response of ZnO NWs may enable these to be used in optical frequency conversion for nanoscale optical circuitry. Semiconductor nanostructures with wide band gaps (e.g., TiO_x) have emerged as important components in a new class of photovoltaic devices. Highly conductive wide band gap transition metal oxide nanostructures have been used in dye-sensitized solar cells to provide a conductive path for the collection of photoexcited carriers. The dye-sensitized solar cells based on aligned ZnO NWs offer features of a large specific surface area, connectivity for carriers, and the desired optical response [16].

References

1. Lopez AM, Alonso AM, Prato M (2011) Materials chemistry of fullerene C_{60} derivatives. J Mater Chem 21:1305–1318
2. Yanilkin VV, Gubskaya VP, Morozov VI et al (2003) Electrochemistry of fullerenes and their derivatives. Russ J Electrochem 39:1147–1165
3. Prato M, Maggini M (1998) Fulleropyrrolidines: a family of full-fledged fullerene derivatives. Acc Chem Res 31:519–526
4. Denis PA (2018) On the estimation of the strength of supramolecular complexes of fullerenes. Int J Quantum Chem 25670:1–5
5. Alonso AM, Tagmatarchis N, Prato M (2006) Fullerenes and their derivatives. Nanomaterials handbook. Taylor & Francis Group, LLC, New York, p 40–79
6. Huang Y, Duan X, Wei Q (2001) Directed assembly of one-dimensional nanostructures into functional networks. Science 291:630–633
7. Cao G, Wang Y (2004) One-dimensional nanostructures: nanowires and nanorods. Nanostructures and nanomaterials. Imperial College Press, London, pp 110–172
8. Duan X, Lieber CM (2000) Laser-assisted catalytic growth of single crystal GaN nanowires. J Am Chem Soc 122:188–189
9. Ren Z, Guo Y, Liu CH et al (2013) Hierarchically nanostructured materials for sustainable environmental applications. Front Chem 1:1–22
10. Li J, Wang D, LaPierre RR (2011) Advances in III-V semiconductor nanowires and nanodevices. https://doi.org/10.2174/97816080505291110101
11. Wang ZL (2004) Mechanical properties of nanowires and nanobelts. Dekker encyclopedia of nanoscience and nanotechnology. Marcel Dekker, Inc., New York
12. Liu S, Sun N, Liu M et al (2018) Nanostructured SnSe: synthesis, doping, and thermoelectric properties. J Appl Phys 123:115109–115115
13. Huang Q, Lilley CM, Bode M et al (2008) Electrical properties of Cu nanowires. In: IEEE conference on nanotechnology. https://doi.org/10.1109/nano.2008.163

14. Bauer J, Fleischer F, Breitenstein O et al (2007) Electrical properties of nominally undoped silicon nanowires grown by molecular-beam epitaxy. Appl Phys Lett 90:012105–012108
15. Joyce HJ, Boland JL, Davies CL et al (2016) A review of the electrical properties of semiconductor nanowires: insights gained from terahertz conductivity spectroscopy. Semicond Sci Tech 31:103003–103023
16. Spanier JE (2006) One-dimensional semiconductor and oxide nanostructures. Nanomaterials Handbook. Taylor & Francis Group, LLC, New York, pp 294–327

Chapter 6
Semiconductors, Organic and Hybrid Nanostructures

Abstract Nanostructures vary according to their materials used as organic, inorganic or hybrid structures. The nanostructured material has at least one dimension confinement and has intermediate size between macro and microscale materials. This chapter describes about the semiconductor nanostructures, organic and hybrid nanostructures and its types, also their fabrication techniques explained in detail.

6.1 Semiconductor Nanostructures

6.1.1 Quasi-One-Dimensional Systems

In nanotubes, the screened potential can be expanded into a series in terms of cylindrical harmonics. The behavior of the zero and all other harmonics is qualitatively different. The zero harmonic, i.e., the axially symmetric part of the potential, experiences logarithmically weak screening:

$$V_0(z) \sim e^2/z \ln^2[z/a]$$

(a is the nanotube radius). All higher harmonics are screened in accordance with the dielectric type, and the effective permittivity depends on the harmonic number as:

$$\varepsilon_n = \left(1 + \frac{\pi k_n}{ma * n}\right)^2$$

where κ_n is the polarization operator of electrons at the zero frequency and zero longitudinal momentum corresponding to virtual transition with a change of the azimuthal quantum number by n [1].

T. D. Thangadurai et al., *Nanostructured Materials*, Engineering Materials,
https://doi.org/10.1007/978-3-030-26145-0_6

6.1.2 Double Quantum Well

The matrix elements of the bare and screened potentials are related by a matrix dielectric function. Its form follows from the system at zero frequency. In the case of a two-component plasma (two parallel plasma layers), the potential of the Coulomb centrum located in one of the layers decreases at large distances as $1/\rho^3$, i.e., in the same manner as it happens in a single layer of two-dimensional electrons. However, the coefficient at ρ^{-3} now depends on the population of the layers, and it is no longer a universal constant $e^{\sim 2}a^{*2}$.

6.1.3 The Size of Semiconductor Nanostructures

The physics of semiconductor nanostructures has a lot in common with other areas of physics. Figure 6.1 is an attempt to illustrate some of these links. The relations with materials science and electronics have already been mentioned above. Beyond that, modern semiconductor electronics is an integrated part of measurement equipment that is being used for the measurement of the physical phenomena. The physics of low temperatures is very important for experimental apparatus such as cryostats which are necessary to reveal quantum phenomena in semiconductor nanostructures. The world of nanostructures starts below a characteristic length of about 1 μm and ends at about 1 nm. Of course, these limits are not strict and not always will all dimensions of a nanostructure be within this interval. For example, a ring with a diameter of 5 μm and a thickness of 300 nm would certainly still be called a nanostructure. The special property of structures within this size range is that typically a few length

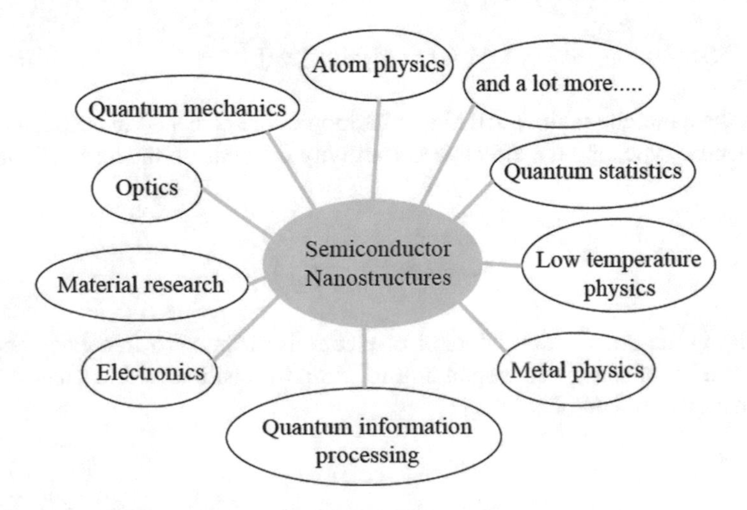

Fig. 6.1 The physics of semiconductor nanostructures is related to many other areas of physics

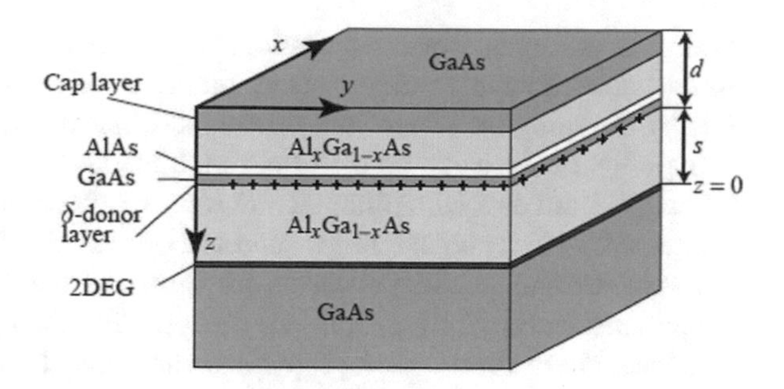

Fig. 6.2 Layer sequence in a typical GaAs/AlGaAs heterostructure with remote doping

scales important for the physics of these systems are of comparable magnitude. In semiconductor nanostructures this could, for example, be the mean free path for electrons, the structure size, and the phase-coherence length of the electrons [2].

6.1.4 Electrostatics of a GaAs/AlGaAs Heterostructure

Consider a GaAs/AlGaAs heterostructure as it is depicted schematically in Fig. 6.2. As a first step we are interested in the electrostatic description of this structure. For simplicity we assume that the relative dielectric constants of GaAs and AlGaAs are identical. We choose the z axis in the growth direction of the crystal, normal to the hetero interface with its origin, $z = 0$, at this interface. The AlGaAs barrier material is in the region $z < 0$, GaAs fills the half space $z > 0$. On top of the GaAs cap layer a thick metal layer has been deposited.

6.1.5 Applications of Semiconductor Nanostructures

The applications of semiconductor nanocrystals in biotechnology have been highlighted recently by a broad variety of applications in the study of subcellular processes of fundamental importance in biology. These applications include the use quantum dots as a new type of fluorescent probes as well as their use as active electronic and optical components in nanostructure–biomolecule complexes of potential utility in influencing bimolecular processes in cells.

Likewise, carbon nanotubes (CNTs) portend many applications in nanobiotechnology as a result of their nanoscale diameters and as a result of the fact that they may be produced as metallic or semiconducting nanostructures. The potential applications of carbon nanotubes in bioengineering were reviewed by its basic structural and electronic properties of carbon nanotubes. Advances in nanomaterials have produced a new class of fluorescent probes by conjugating semiconductor quantum dots (QDs), also known as semiconducting nanocrystals, with biomolecules that have affinities for binding with selected biological structures. These inorganic dyes have great advantages over conventional organic dyes; such advantages include the option for continuously and precisely tuning the emission wavelength of quantum dots by changing the size of the nanocrystal; narrow symmetric emission spectra; a single light source can be used for simultaneous excitation of multiple semiconductor quantum dots with different emission spectra of longer wavelengths than the source; ability to function through repeated cycles of excitation and fluorescence for hours; and extreme stability of coated quantum dots against photobleaching as well as changes in the pH of the biological electrolytes that are ubiquitous in biological environments. These novel optical properties render quantum dots ideal fluorophores for ultrasensitive, multicolor, and multiplexing applications in cell and molecular biology as well as in bioengineering. Similar studies with carbon nanotubes (CNTs) have emphasized the use of chemically functionalized CNTs to achieve specificity in binding to biological structures and how an electrolyte facilitates the control of the electrical properties of a CNT [3].

6.2 Organic Nanostructures

Organic electronic material's has made remarkable technological breakthroughs for the last decade, enabling the realization of viable devices such as organic light-emitting diodes (OLEDs), organic field-effect transistors (OFETs), organic solar cells as well as memory circuits. Some products such as medium resolution emissive OLED displays are already commercially available and others are now in various stages of commercialization. On the other hand, nanoscale technology has clear advantages to be gained from exploiting self-organized growth, as it avoids the need for highly sophisticated patterning of surfaces with nanometer size objects.

Preferably, in organic applications the functional properties can be obtained essentially in one self-assembled molecular layer, so that organic electronics in principle would offer a maximum degree of miniaturization. Aniosotropic molecule especially p-sexiphenyl shows a typhical solid state herring bone structure (Fig. 6.3). π–conjugated polymers combine properties of classical semiconductors with the inherent processing advantages of plastics and therefore play a major role in low cost, large area optoelectronic applications. Unfortunately, polymers are highly disordered in the solid state. Consequently, carrier transport is dominantly influenced by localization resulting in low charge carrier mobilities ($\mu \ll 1\ \mathrm{cm^2\ V^{-1}\ s^{-1}}$). Therefore, an important part of research aims toward significant improvement in the performance of organic devices and deeper understanding of physical processes using small molecular systems, in which highly ordered or even crystalline states can be obtained. Two different approaches towards well-ordered small molecular systems are distinguished, namely (1) the growth of free-standing single crystals of high purity and (2) thin film growth by epitaxial vacuum preparation techniques. Unfortunately, bulk organic single crystals showing very high mobilities are basically technologically irrelevant because of their poor mechanical stability. Therefore, there is a great interest in the second alternative, which can lead to highly ordered self-organized films with well-defined orientation of the molecules [4].

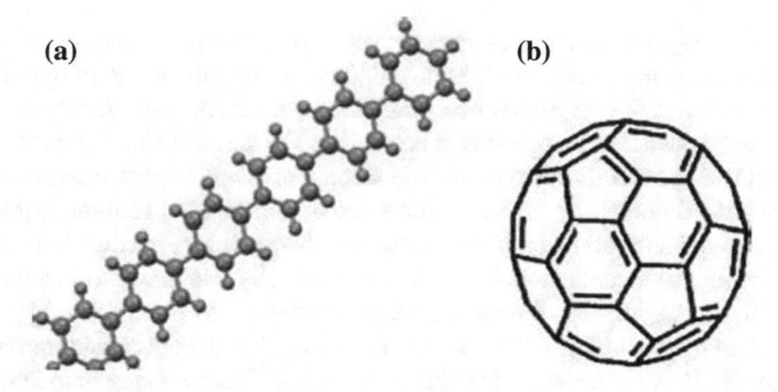

Fig. 6.3 Schematic of para-sexiphenyl ρ-sexiphenyl $C_{36}H_{26}$ **(a)** and C_{60} **(b)** molecules indicating single and double C–C bonds

6.2.1 Structures and Applications of Organic Nanostructures

In contrast to oligo-phenylenes, fullerene C_{60} represents the class of highly symmetrical molecules. The closed-cage nearly spherical molecule C_{60} and related fullerene molecules (C_{70}, C_{84}, C_{60} derivatives, etc.) have attracted a great deal of interest in recent years because of their unique structure and properties. In particular, upon photoexcitation C_{60} is known as excellent electron acceptor relative to conjugated polymers and oligomers, capable of taking on as many as six electrons. In the case of donor–acceptor (D–A) blends or bilayers, this leads to ultrafast photoinduced charge transfer with subsequent long-lived charge separated states. Such D–A systems were successfully used in fabricating organic solar cells and photosensitive OFETs. On the other hand, C_{60} is a very good material for highly ordered thin films, showing well-aligned epitaxial growth with the {111} planes of C_{60} parallel to the surface of layered substrates and especially of mica [5, 6].

6.2.2 Miscellaneous Application of Organic Nanostructures

Due to the prospective performances in photonic integrations, one-dimensional single crystalline nanostructures constructed from organic luminescent molecules have generated wide research interests like: Inspired by the existing optically and electrically pumped inorganic semiconductor and polymer nanowire lasers, researchers show increasing interest in organic-based lasers, for their unique features, such as a high degree of spectral tunability, large stimulated emission cross-sections, and the potential for simple high-throughput fabrication. Moreover, the ease of fabrication has made it possible to produce 1D organic laser with a wide variety of optical microcavities. Therefore, the fabrication of 1D organic nanolasers is of great scientific interest and technological significance.

Optical waveguides can lead to some properties of laser light, because waveguides have common laser geometries. 1D waveguiding nanostructures with optically flat end facets can function as miniature optical cavities, which apply feedback. As the light passes backwards and forwards, it is amplified by the stimulated emission, and if the amplification exceeds the losses of the resonator, lasing begins. Nanolasers with strong polarized output are able to offer great opportunities in future applications such as nanooptical routing, emission, detection, data storage, sensing and near-field optics. Organic semiconductors have been successfully adopted as optically driven lasers, while direct electrical pumping of organic lasers is a very challenging problem, mostly relating to the low mobilities of the materials. However, in consideration of the advances in molecular design and synthesis, as well as the development in fabricating single-crystalline organic nanowires, which exhibit excellent performances of photon confinement and propagation and high charge-carrier mobilities, it is reasonable to consider that organic lasers show great promise in the future of electrically pumped laser devices [7].

6.3 Hybrid Nanostructures

The hybrid nanostructures of metal NPs/2D materials are commonly used in SERS and photocatalytic reactions, and as the interlayers of solar cells. There are certain preparation requirements for specific applications, such as a large surface area for SERS and photocatalytic reactions, and solution processing for the interlayers of solar cells.

6.3.1 Physical Deposition to Synthesize the Hybrid Nanostructures of Metal NPs/2D Materials

Physical deposition is a convenient method that can rapidly and efficiently deposit metal NPs on 2D nanosheets. Sputter deposition, electron-beam evaporation, and thermal evaporation are three main physical deposition methods. 2D materials need to be synthesized first, followed by conversion of the bulk metal into small metal NPs in a vacuum or an inert atmosphere to construct the hybrid nanostructures of metal NPs/2D materials. Hybrids of metal NPs/2D materials with different densities, shapes, and thicknesses of metal NPs can be obtained using this method, which is very beneficial for tuning the plasmonic effect. The deposition of Au or Ag NPs on GO was patterned using sputter deposition technique. Sometimes, an annealing treatment is needed to obtain nanospheres. For example, by adjusting the annealing temperature and time, Au nanospheres can be obtained on graphene sheets. The Ag NPs/graphene composite nanostructures fabrication using thermal evaporation was made, in that sandwiched graphene hybrid nanostructures were fabricated by transferring a monolayer of graphene on top of an Ag film, followed by placing Ag NPs on the graphene surface through thermal evaporation [8].

6.3.2 Chemical Reduction to Synthesize the Hybrid Nanostructures of Metal NPs/2D Materials

Chemical reduction is the simplest and most efficient method for anchoring metal NPs onto 2D materials with the assistance of various chemical reagents in a solution. From a broad point of view, the hybrid nanostructures of metal NPs/2D materials can usually be described as various metal NPs (Au, Ag, Cu, etc.) with uniform or different shapes that are decorated on graphene or other 2D materials. Technically, the hybrid nanostructures of metal NPs/graphene are prepared via two routes: (1) mixing and reducing either the metal precursors or GO precursors and (2) simultaneously mixing and reducing the metal precursors and GO precursors.

6.3.3 Applications of Hybrid Nanostructures

A plasmon-enhanced laser is also an important research direction. By transferring graphene to the top of an Au film absorber mirror, stable mode-locking bulk lasers with wavelengths of 1, 2, and 2.4 mm have been fabricated. Plasmon enhanced near-infrared absorption in graphene with metal gratings. In addition, the hybrid Ag nanowire/graphene nanostructure can be used as a highly transparent and stretchable FET sensor for detecting some biomolecules.

The hybrid system of metal NP/graphene also shows an amplified photoacoustic effect, suggesting its implementation as an imaging probe. Based on the enhanced near-infrared absorption and enhanced photothermal stability in the Au nanorod/RGO composite, an extraordinarily high photoacoustic effect in the 4–11 MHz operating frequency of an ultrasound transducer has been demonstrated in in vitro and in vivo systems. Under light irradiation, especially near-infrared light, metal nanomaterials can produce thermal energy via nonradiative decay. In addition, 2D materials such as RGO can also absorb light in the broadband region ranging from UV to near-infrared wavelengths. Based on its nonradiative decay, GO can also serve as a photothermal source. By combining metal NPs and GO, the metal NPs enhance the light absorption of GO via the plasmonic near-field effect, and simultaneously, improved infrared absorption and photothermal effects are produced by the hybrid system. A strong photothermal effect has been achieved in a GO coated Au nanorod hybrid system, which shows promise for biomedical applications, such as killing cells [9].

References

1. Latyshev AV, Dvurechenskii AV, Aseev AL (2017) Advances in semiconductor nanostructures, growth, characterization, properties and applications. Elsevier, Amsterdam
2. Majumdar A (2004) Thermoelectricity in semiconductor nanostructures. Science 303:777–778
3. Alexson D, Chen H, Cho M et al (2005) Semiconductor nanostructures in biological applications. J Phys: Condens Matter 17:R637–R656
4. Park JG, Kim GT, Park JH et al (2001) Quantum transport in low dimensional organic nanostructures. Thin Solid Films 393:161–167
5. Atwood JL, Steed JW (2008) Organic nanostructures. Wiley, Weinheim
6. Andreev A, Teichert C, Singh B et al (2008) Fabrication and characterization of self-organized nanostructured organic thin films and devices. Organic nanostructures for next generation devices. Springer, Heidelberg, pp 263–300
7. Cui QH, Zhao YS, Yao J (2012) Photonic applications of one-dimensional organic single-crystalline nanostructures: optical waveguides and optically pumped lasers. J Mater Chem 22:4136–4140
8. Lu G, Li H, Liusman C et al (2011) Surface enhanced Raman scattering of Ag or Au nanoparticle-decorated reduced graphene oxide for detection of aromatic molecules. Chem Sci 2:1817–1821
9. Li X, Zhu J, Wei B (2016) Hybrid nanostructures of metal/two-dimensional nanomaterials for plasmon-enhanced applications. Chem Soc Rev 45:3145–3187

Chapter 7
Properties of Nanostructured Materials

Abstract Nanostructured materials are becoming of major significance, and their investigations require a comprehensive approach. Depending on the size of the smallest feature, the interaction of light with structured materials can be very different. The chapter identifies the nanostructures unique properties that make the nanostructure for varying applications and detailed about the properties.

7.1 Unique Properties of Nanostructures

The buckyballs and buckytubes have sincerely significant properties: especially their high electrical conductivity and unique mechanical strength. The properties of surfactant-stabilized colloids are the basis for many bioanalytical systems. Fluorescent CdSe quantum dots, unlike fluorescent organic dyes, photobleach only slowly, and show interesting optical phenomena such as blinking [1].

Quantum behavior becomes increasingly prominent as structures become smaller. Many of the behaviors of atoms and molecules are, of course, only explicable on the basis of quantum mechanics. The properties of objects and structures larger than a few microns are usually classical. In the intermediate region-the region of nanometer-scale structures quantum and classical behaviors mix. This mixture offers the promise of new phenomena and/or new technologies. The fluorescent behavior of semiconductor quantum dots can only be explained quantum mechanically; as can the tunneling currents that characterize scanning tunneling microscopes, and electron emission from the tips of buckytubes. The response of electrical resistance to magnetic field in giant magnetic resonance (GMR) materials is already useful in magnetic information storage, and the behavior of spin-polarized electrons in magnetic semiconductors forms one foundation for the emerging field of spintronics. The ability to make structures in the region where quantum behavior emerges, or where classical and quantum behaviors merge in new ways, is one with enormous opportunity for discovery. And because quantum behavior is fundamentally counter-intuitive, there is the optimistic expectation that nanostructures and nanostructured materials will found fundamentally new technologies. The ability to fabricate small cantilevers, tips, and wires, opens the possibility of making nanoscale sensors [2].

T. D. Thangadurai et al., *Nanostructured Materials*, Engineering Materials,
https://doi.org/10.1007/978-3-030-26145-0_7

7.2 Physical Properties of Nanowires

7.2.1 Thermal Stability

While Comparing with bulk materials, the low-dimensional nanoscale materials, with their large surface area and possible quantum confinement effect, exhibit distinct electric, optical, chemical and thermal properties. Thermal stability of the semiconductor nanowires is of critical importance for their potential implementation as building blocks for the nanoscale electronics. Size-dependent melting recrystallization process of the carbon-sheathed semiconductor Ge nanowires has been recently studied by the Yang group using an in situ high temperature transmission electron microscope.

The silicon-based nanostructures have different morphologies and microstructures at different formation and annealing temperature. The synthetic method is thermal evaporation of Si/SiO_2 mixture in an alumina tube. It was observed that besides Si nanowires, many other kinds of Si based nanostructures such as octopuslike, pinlike, tadpolelike, and chainlike structures were also formed. The formation and annealing temperature was found to play a dominant role on the formation of these structures. It is demonstrated that a control over the temperature can precisely control the morphologies and intrinsic structures of the silicon-based nanomaterials. This is an important step toward design and control of nanostructures using the knowledge of nanowire thermal stability [3].

7.2.2 Optical Properties

7.2.2.1 Photoluminescence and Stimulated Emission

Due to quantum confinement effect, the nanowires exhibit distinct optical properties when their size is below certain critical dimension. For example, the absorption edge of the Si nanowires synthesized and shows blue-shift from the bulk indirect band gap of 1.1 eV. Sharp discrete absorbance features and relatively strong band edge photoluminescence (PL) were also observed. These optical properties likely result from quantum confinement effects, although they cannot rule out the possibility of additional surface states as well. The <110> oriented nanowires exhibited distinctly molecular-type transitions. The <100> oriented wires exhibited a significantly higher exciton energy than the <110> oriented wires, are among the first to show that the tunability of the lattice orientation in the silicon nanowires can lead to different optical properties [4].

7.2.3 Electronic Properties

Miniaturization in electronics through improvements in established top-down fabrication techniques is approaching the point where fundamental issues are expected to limit the dramatic increases in computing speed. Semiconductor nanowires have recently been used as building blocks for assembling a range of nanodevices including FETs, p-n diodes, bipolar junction transistors, and complementary inverters. In contrast to carbon nanotubes, the nanowire devices can be assembled in a predictable manner because the electronic properties and sizes of the nanowires can be precisely controlled during synthesis and methods are being developed for their parallel assembly. Researchers explored the possibility of assembling various devices at the nanometer scale using their n- and p-type Si, InP nanowires. It is believed that bottom-up approach to nanoelectronics has the potential to go beyond the limits of the traditional top-down manufacturing techniques. As the critical dimension of an individual device becomes smaller and smaller, the electron transport properties of their components become an important issue to study. As for semiconductors, recent measurements on a set of nanoscale electronic devices indicated that GaN nanowires as thin as 17.6 nm could still function properly as a semiconductor. Another issue related to the electronic applications of chemically synthesized nanowires is the assembly of these building blocks into various device architectures. It is worth noting that the Lieber group has been able to assemble semiconductor nanowires into cross-bar p-n junctions and junction arrays having controllable electrical characteristics with a yield as high as 95%. These junctions have been further used to create integrated nanoscale field-effect transistor arrays with nanowires as both the conducting channel and gating electrode. In addition, OR, AND, and NOR logic-gate structures with substantial gain have been configured and tested to implement some basic computation. There are several appealing features for bottom-up approach to nanoelectronics. First, the size of the nanowire building blocks can be readily tuned to sub-100 nm and smaller, which should lead to high density of devices on a chip. Second, the material systems for the nanowires are essentially unlimited, which should give researchers great flexibility to select the right materials for the desired device functionality. For example, the GaN nanowire based nanodevices which would be of interest for their high power/high temperature electrical applications [5].

7.2.4 Mechanical Properties

The mechanical properties of small, rodlike materials are of considerable interest. For example, small whiskers can have strengths considerably greater than those observed in corresponding macroscopic single crystals, an effect that is attributed to a reduction in the number of structural defects per unit length that lead to mechanical failure. The researchers used atomic force microscopy to determine the mechanical properties of individual, structurally isolated silicon carbide (SiC) nanowire that were pinned at one end to molybdenum disulfide surfaces. The bending force was measured versus displacement along the unpinned lengths. Continued bending of the SiC nanowires ultimately led to fracture. They calculated Young's modulus of 610–660 Gpa based on their AFM measurement. These results agree well with the 600 GPa value predicted theoretically for [111]-oriented SiC and the average values obtained previously for micrometer-diameter whiskers. The large Young's modulus values determined for SiC nanowires make these materials obvious candidates for the reinforcing element in ceramic, metal, and polymer matrix composites [6].

7.2.5 Field Emission Properties

It is well-known that nanotubes and nanowires with sharp tips are promising materials for applications as cold cathode field emission devices. Field emission characteristics of the β-SiC and Si nanowires have been investigated using current-voltage measurements. The silicon carbide nanorods exhibited high electron field emission with high stability. Both Si and SiC nanowires exhibit well-behaved and robust field emission. The turn-on fields for Si and SiC nanowires were 15 and 20 V μm^{-1}, respectively and current density of 0.01 mA cm^{-2} which are comparable with those for other field emitters including carbon nanotubes and diamond. Along with the ease of preparation, these silicon carbide and Si nanowires are believed to have potential application in electron field emitting devices [7].

7.3 Grain Boundaries in Nanostructured Materials

With grain sizes in a submicron (100–1000 nm) or nanocrystalline (<100 nm) range, (Ultrafine-grained materials) UFG materials contain in their microstructure a very high density of grain boundaries, which can play a significant role in the development and exhibition of novel properties. For this reason, UFG materials can be typically considered as interface-controlled materials. In pioneered nanocrystalline materials it was suggested that grain boundaries can possess a number of peculiar features in terms of their atomic structure in contrast to grain boundaries in conventional polycrystalline materials. This is of particular importance for UFG materials produced by (Severe Plastic Deformation) SPD methods. Depending on the regimes of SPD processing, different types of grain boundaries can be formed in the UFG materials (high- and low-angle, special and random, and equilibrium and so-called non-equilibrium grain boundaries containing extrinsic dislocations), which paves the way to grain boundary engineering of UFG materials, i.e., to the control of their properties by means of varying the grain boundary structure. The notions on non-equilibrium grain boundaries were first introduced in the scientific literature in the 1980s, reasoning from investigations of interactions of lattice dislocations with grain boundaries.

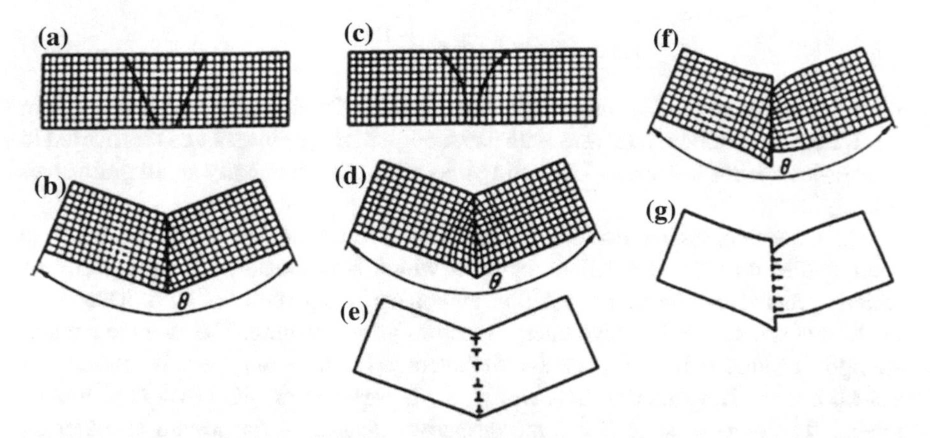

Fig. 7.1 Phenomenological engineering of grain boundaries via thought cuts: (**a**) → (**b**) equilibrium grain boundary (bunches of deformation are connected without mesoscopic strain); (**c**) → (**d**) and (**a**) → (**f**) non-equilibrium grain boundaries (deformation is required for crystal joint – bending and tension – compression, respectively); (**e**) and (**g**) schemes of grain boundary dislocation (GBD) complexes initiating the same character of mesoscopic elastic distortions as in (**d**) and (**f**) Reprinted with permission from © In: Bulk Nanostructured Materials with Multifunctional Properties. Springer Briefs in Materials. Springer, Cham (2015)

According to, the formation of a non-equilibrium grain boundary state is characterized by three main features, namely excess grain boundary energy, the presence of long-range elastic stresses, and enhanced free volume. Discontinuous distortions of crystallographically ordered structures, which may come about by accommodation problems of differently oriented crystallites of finite sizes or by high densities of lattice dislocations and their interaction with grain boundaries, can be considered as sources of elastic stress fields that modify the atomic structure of high-angle grain boundaries so that their excess free energy becomes enhanced. These unusual grain boundaries are known as non-equilibrium grain boundaries, each grain boundary is non-equilibrium defect if segregation effects are not to be considered [8] (Fig. 7.1).

7.4 Multifunctional Properties of Nanostructured Metallic Materials

7.4.1 Mechanical Properties

The mechanical and functional properties of all polycrystalline metallic materials are determined by several factors, the grain size of the material generally plays the most significant and often a dominant role. Thus, the strength of different polycrystalline materials is related to the grain size, d, through the Hall–Petch equation which states that the yield stress, σ_y, is given by

$$\sigma_y = \sigma_o + k_y d^{-1/2} \quad (1.1)$$

where σ_o is termed the friction stress and k_y the Hall–Petch constant. It follows from Eq. 1.1 that the strength increases with a reduction in the grain size and this has led to an ever-increasing interest in fabricating materials with extremely small grain sizes [9].

The mechanical properties like flexural strength and the fracture toughness have been studied on a sintered silicon carbide, which is prepared by the pressure less sintering route from the nano crystalline silicon carbide particles of an Acheson type α-SiC that is processed by high energy attrition grinding route. The average flexural strength is found to be 390 MPa and the average fracture toughness is found to be 4.3 MPa $m^{1/2}$. The sintering behavior of nano particles of SiC is very interesting indeed. The non-oxide ceramics are attractive candidates for structural materials because of their high temperature strength, which makes them highly suitable for the applications in ceramic engines and gas turbines.

The silicon carbide seems to be particularly well suited because of its high temperature strength and intrinsic resistance to oxidation. However, silicon carbide ceramic has a strong covalent bonding character and a small amount of sintering aids such as AlN or B_4C are usually necessary to make dense materials. Usually, the dense silicon carbide materials have lower strength compared to silicon nitride materials. The latter experience of a degradation of strength at 1200–1300 °C, while silicon carbide materials do not show any decrease of strength up to 1500 °C [10]. The silicon carbide, like other ceramics, shows a wide scattered distribution in flexural strength. This distribution of strength is a matter of concern and hence an important topic to be studied. The strength distribution of silicon carbide varies typical for many other ceramic materials. Most of earlier work is based on a 4-point bending test. Depending on the manufacturer and the type of silicon carbide, the average room temperature strength varies between 350 and 550 MPa. For such cases, the Weibull modulus which is a measure of the dispersion in strength varies between 6 and 15 depending on the strength of the ceramic materials. The observed behavior in metals, the room temperature strength of the sintered α-silicon carbide increases, as the exaggerated grain growth is intentionally introduced by a suitable heat treatment. However, the comparisons between the grain sizes should be made between two different materials having a uniform distribution of small grain sizes versus a uniform distribution of large grain sizes. The isolated rise of a mixed grain-size matrix, it should be pointed out that the mechanical strength of the sintered silicon carbide remains unaltered at temperatures up to 1500 °C, which has a definite advantage in the high temperature applications such as gas turbines and ceramic engines. A further increase of strength has been observed in this material, when heated in the argon atmosphere the effect of the dopants on the flexural strength. They observed that the materials doped with aluminum had a higher strength than that doped with boron, and with the improvement in the processing parameters, it would be as strong as the 'hot-pressed' SiC materials. They also found that the production of the materials with ultrafine grains occurred in a very narrow range of sintering temperature (2050–2075 °C). The exaggerated grain growth at temperatures above 2075 °C was the cause of the reduction in strength of boron doped materials. The materials sintered with aluminum nitride, which acted as a grain growth inhibitor, showed a higher strength. The silicon carbide, like other ceramics, shows a wide distribution in strength.

One common way of characterizing the variability of strength of brittle materials is by using the Weibull modulus. A high Weibull modulus material (i.e. $m > 20$) will give a narrow distribution of fracture strength, whereas a low Weibull modulus material (i.e. $m < 5$) has a wide distribution in fracture strength, and hence a low reliability. However, a number of workers indicated that the structural reliability of ceramics is primarily dependent on the flaw size distribution during the processing [11].

7.4.2 *Strength Measurement*

7.4.2.1 Flexural Strength

The measurement of flexural strength requires the following steps:

1. Preparation of the Samples,
2. Introduction of 4-Point Bending Samples, and
3. Measurement of the Fracture Load.

7.4.2.2 Fracture Toughness

A large number of methods of measurement of fracture toughness and such other parameters have been standardized for metals and alloys in recent years. These methods are normally suitable for ambient laboratory conditions. Different methods utilize different configurations. In sintered nano-crystalline particles of silicon carbide, the measurement of fracture toughness was made by 4-point Single Edge Notch Bend (SENB) specimen technique. The load required for causing the fracture of the sample was measured by the above high temperature Bending Strength Tester (BST), having a length 1600 mm × width 750 mm × height 2100 mm. The bars of size (45 mm × 4.5 mm × 3.5 mm) were notched with a diamond blade of 0.2 mm thickness (100 μm diameter), after grinding in a 25 μ diamond wheel. The notch dimensions were measured with the help of a traveling microscope. As a result of using 0.2 μm thick blade for notching, the notch width varied from 0.31 to 0.39. The value of a/w ≈ 0.25–0.40 mm, where a = notch length and w = notch width. The notched samples were put into the furnace. On attainment of the desired temperature, the load was applied on the Bending Strength Tester at the speed of 1.25 N/s, and the fracture load was noted [12].

7.4.2.3 Flexural Strength of α-SiC

There are many factors which affect the strength of ceramic materials such as porosity, grain size, grains shape and surface conditions etc. Figure 7.2 shows a SEM micrograph of a polished and etched surface of the dense α-silicon carbide. The grains had mostly tabular shapes. The average grain size was found to be 5700 nm. Figure 7.3 shows the flexural strength as a function of temperature. No variation of strength data with temperatures was found up to 1400 °C. The average flexural strength was found to be 390 MPa. Figures 7.3 and 7.4 show the flexural strength

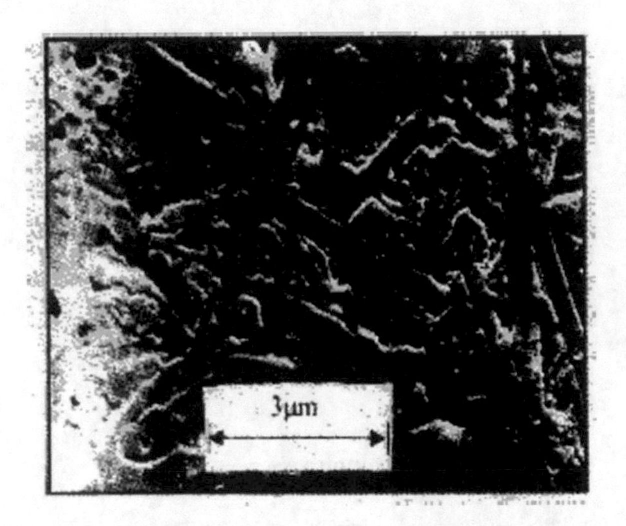

Fig. 7.2 SEM photo of etched and polished surface of a dense α-silicon carbide (dopant: 0.5 wt% boron carbide + 1 wt% carbon)

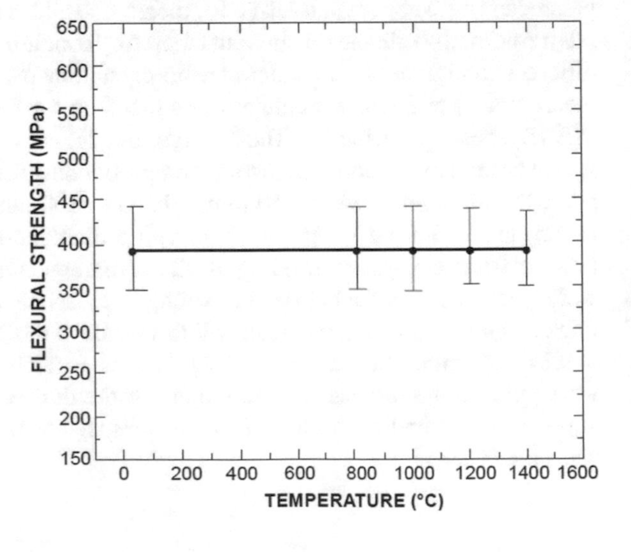

Fig. 7.3 Flexural strength as a function of temperature of α-silicon carbide (dopant: 0.5 wt% boron carbide + 1 wt% carbon)

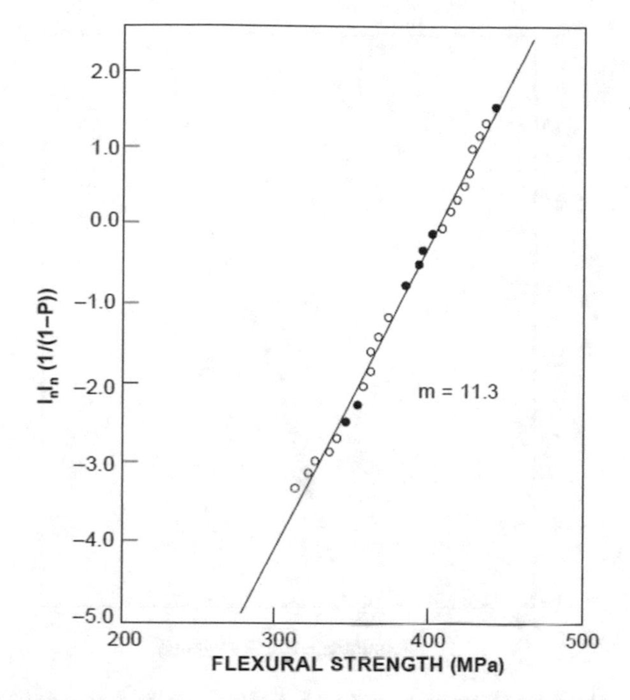

Fig. 7.4 Weibull modulus of flexural strength at room temperature for α-silicon carbide (dopant: 0.5 wt% boron carbide + 1 wt% carbon)

as Weibull modulus (m) at room temperature and at 1400 °C. The number of specimens was 30 in each case. The Weibull modulus was 11.3 at room temperature and 13.3 at 1400 °C. The Weibull modulus is found to increase at higher temperature presumably due to healing of surface cracks, which were induced during the cutting of the samples by high speed diamond blades, i.e. the curing of the surface damages at higher temperature. The Weibull modulus for β-silicon carbide (not shown here) was found to be 10.9 and 12.9 at room temperature and 1400 °C respectively. The low value of 'm' for β-silicon carbide could be due to the presence of pores of different sizes and also due to a relatively wider distribution of flexural strength at room temperature. The increase of Weibull modulus by about 20% for both these materials from room temperature to 1400 °C might be due to the 'healing of surface cracks' and possibly due to the 'release of residual stresses'. Figure 7.6 shows the fracture toughness of α-silicon carbide as a function of temperature. The average fracture toughness was found as 3.95 MPa $m^{1/2}$. The value for β-silicon carbide was found to be 3.55 MPa $m^{1/2}$. A higher value of fracture toughness for α-silicon carbide is obtained due to the fine grained microstructures obtained after sintering and due to higher level of densification compared to that of β-silicon carbide [13] (Fig. 7.5).

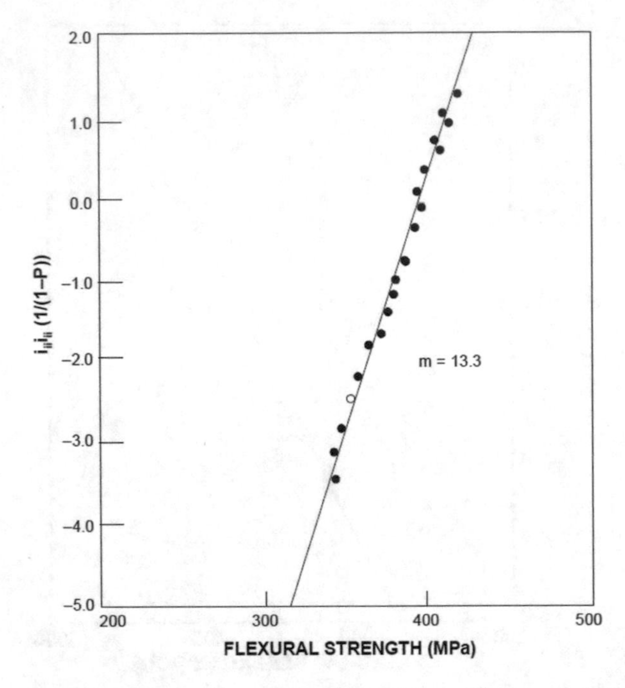

Fig. 7.5 Weibull modulus of flexural strength at 1400 °C for α-silicon carbide (dopant: 0.5 wt% boron carbide + 1 wt% carbon)

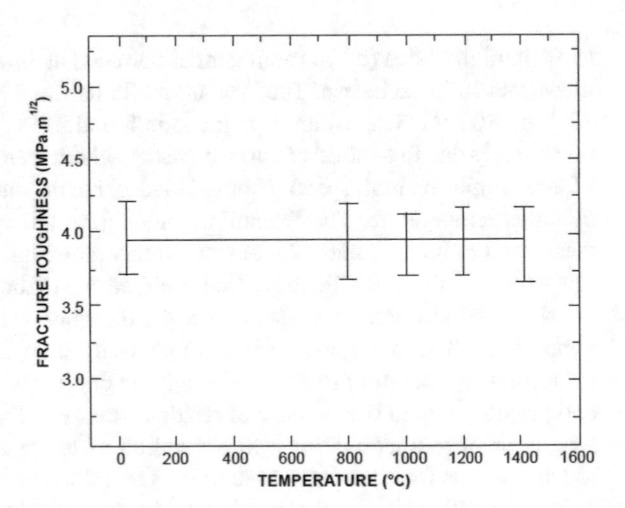

Fig. 7.6 Fracture toughness as a function of temperature of α-silicon carbide (dopant: 0.5 wt% boron carbide + 1 wt% carbon)

7.4.2.4 Microstructures

The fracture mainly originates from the crack generated due to pores. This type of fracture, which is caused by an intergranular pore, was observed during this investigation. Figure 7.7 shows a SEM micrograph of the fracture surface of the α-silicon

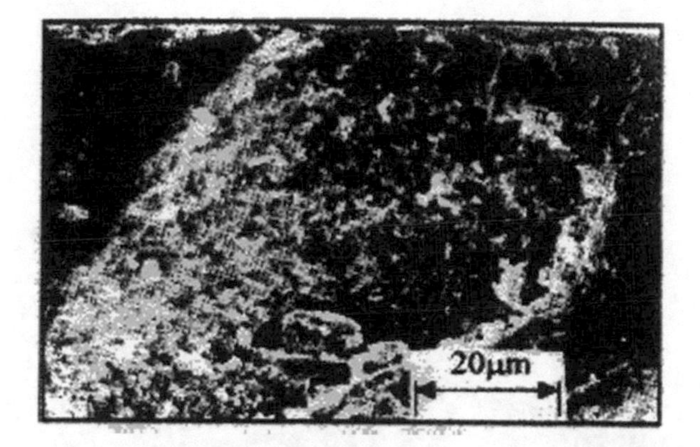

Fig. 7.7 SEM photo showing the presence of a large void (the origin of the crack)

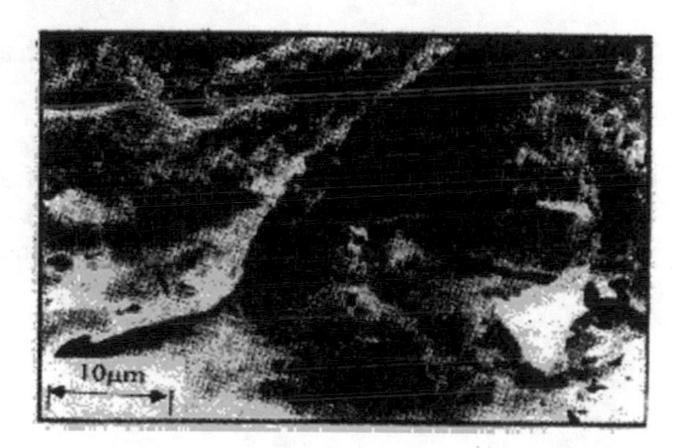

Fig. 7.8 SEM photo showing crack propagation through the sintered sample

carbide in terms of the propagation of cracks and the fracture caused by the inter-granular pores. The presence of pores is expected to cause the formation of tiny cracks. The propagation of cracks requires the rupture of the inter-atomic bonds at the tip of the crack [14] (Fig. 7.8).

7.4.3 Superstrength and Ductility

The influence of grain size reduction to nanometer range on the materials strength has been explored in multiple works. The enhancement of strength with grain size reduction in compliance was observed in many studies, but for nanosized grains (20–50 nm), this relations reported to be typically violated so that the Hall–Petch plot deviates from linear dependence at lower stress values and its slope k_y often becomes negative (the curve 1 in Fig. 7.9). The Hall–Petch relationship breakdown is not observed in ultrafine-grained materials with a mean grain size of 100–1000 nm

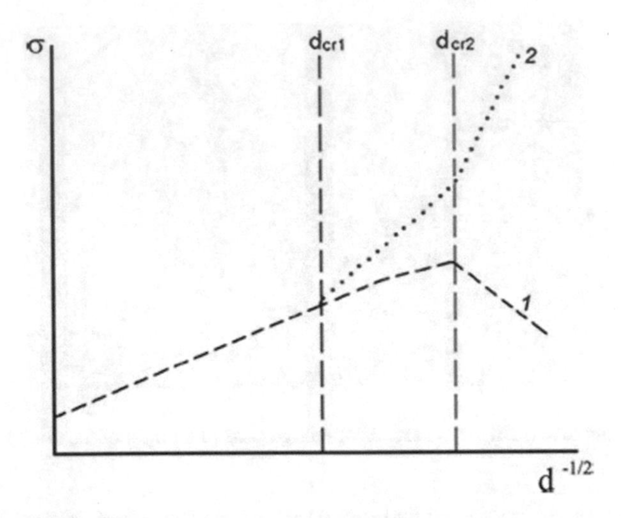

Fig. 7.9 The Hall–Petch slopes within different characteristic length scales

produced by SPD processing. Moreover, recently it was shown that UFG alloys can exhibit a considerably higher strength than the Hall–Petch relationship predicts for the range of ultrafine grains. The nature of such superstrength may be associated with another nanostructural features which could be observed in the SPD-processed metals and alloys [15].

7.4.4 Electrical Conductivity

Electrical conductivity is an important functional property of metallic materials. Electrical conductivity in metals is a result of the movement of electrically charged particles. The atoms of metal elements are characterized by the presence of valence electrons—electrons in the outer shell of an atom that are free to move about. These free electrons allow metals to conduct an electric current. Because valence electrons are free to move, they can travel through the lattice that forms the physical structure of a metal. Under an electric field, free electrons move through the metal much like billiard balls knocking against each other, passing an electric charge as they move. The transfer of energy is strongest when there is little resistance. On a billiard table, this occurs when a ball strikes against another single ball, passing most of its energy onto the next ball. If a single ball strikes multiple other balls, each of those will carry only a fraction of the energy. By the same token, the most effective conductors of electricity are metals that have a single valence electron that is free to move and causes a strong repelling reaction in other electrons. This is the case in the most conductive metals, such as Ag, Cu, and Au, which have a single valence electron that moves with little resistance and causes a strong repelling reaction. Al with three valence electrons comes as the fourth most conductive metal. Mechanical

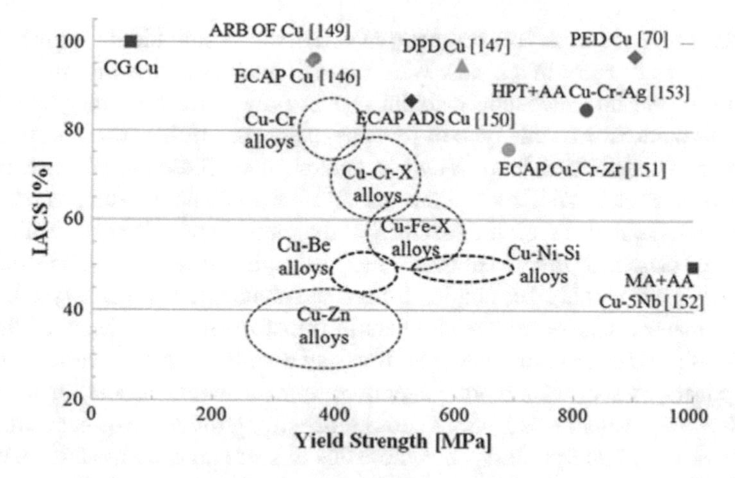

Fig. 7.10 Room temperature electrical conductivity (in IACS) versus yield strength for CG pure Cu and Cu-based alloys and their nanostructured counterparts obtained via SPD processed Reprinted with permission from © Journal of Minerals, 64, 1134–1142 (2012)

strength and electrical conductivity are the most important properties of conducting materials. However, high strength and high electrical conductivity are mutually exclusive in metallic materials, and trade-off between strength and conductivity is always encountered in developing conducting materials. Pure Cu and Al having high electrical conductivity show very low mechanical strength. The alloying of pure metals, strain hardening, or introduction of precipitates dramatically improves their mechanical strength due to suppression of dislocation glide. But all strengthening methods lead to a degradation of electrical conductivity, since the latter is determined by the scattering of conductive electrons due to disturbances in the crystal structure including thermal vibrations, solute atoms, and crystal defects [16]. This is clearly seen from the electrical conductivity—yield strength plot for coarse-grained pure Cu and its alloys presented in (Fig. 7.10). The electrical conductivity of most conventional Cu alloys ranges from ~30 to ~85% of International Annealed Copper Standard (IACS).

7.4.5 Magnetic Properties

Ferromagnetic materials with the demagnetized state do not show magnetization although they have spontaneous magnetization. This is because the ferromagnetic materials are divided into many magnetic domains. Within the magnetic domains, the direction of magnetic moment is aligned. However, the direction of magnetic moments varies at magnetic domain walls so that it can reduce the magnetostatic energy in the total volume. When domain wall can easily migrate, the ferromagnetic material can be easily magnetized at low magnetic field. This type of ferromagnetic materials is referred to as soft magnetic material and is suitable for applications of magnetic cores or recording heads. When domain wall is difficult to migrate, magnetization of the ferromagnetic material occurs only when high magnetic field is applied. In other words, this type of ferromagnetic materials is difficult to magnetize, but once magnetized, it is difficult to demagnetize. These materials are referred to as hard magnetic materials and are suitable for applications such as permanent magnets and magnetic recording media. Soft magnetic materials are one of the most investigated classes of functional materials. Their microstructure can be amorphous or crystalline with grain sizes ranging from some nanometers in nanocrystalline soft magnetic materials to several centimeters in transformer steels. Most of them are available as powders, ribbons, and thin films and are used in a huge variety of industrial applications such as motors, generators, transformers, sensors, or microelectronic devices. Earlier investigations showed the strong influence of nanostructuring via SPD on magnetic hysteresis characteristics of soft magnetic metals, such as Ni, Co, Fe, and Fe–Si. A sharp increase in the coercivity Hc was revealed in pure metals after SPD processing. It was demonstrated that coercivity in nanostructured pure Ni strongly depends on grain size, dislocation structure, and the non-equilibrium state of the grain boundaries. For example, SPD-processed specimens subjected to recovery annealing at 100 and 200 °C had different values of coercivity by ~ 40%, though their grain sizes were nearly the same. At the same time, at 200°C, an intense recovery of non-equilibrium grain boundaries was observed. As the temperature of annealing increased, a further decrease in the value of Hc was correlated with increasing grain size. An analogous regularity was revealed in Co processed via (high pressure torsion) HPT. TEM investigations of the domain structure of nanostructured Co with a grain size of about 100 nm used the Lorentz method. It was established that the magnetic domain sizes were significantly larger than the grain sizes and remagnetization was conditioned by the movement of domain walls, but at the same time, this movement was hampered by non-equilibrium grain boundaries [17].

7.4.6 Corrosion Resistance

Corrosion, the property of a material to deteriorate its properties due to interaction with the environment, especially chemically aggressive one, is an old problem for the industry since the very beginning. The embrittlement of a metal, leading to cracking of metallic materials inducing catastrophic failures of constructions, inspired the challenge to modify microstructures to survive hostile conditions. Materials required by different branches of industry are generally described by corresponding standards, but they are not flexible enough to reflect the industry needs. Development of corrosion-resistant metallic materials for advanced applications is a topical task for modern materials science with attention to both effectiveness and security of the cutting-edge technologies and ecological security taking into consideration risks of global warming and industrial disasters. The problem of utmost importance is development of new functional materials with reasonable corrosion resistance to be used in new-generation applications. Corrosion resistance was shown to be significantly enhanced in several nanocrystalline metals and alloys as compared to their coarse-grained counterparts. A tendency for localized corrosion was observed to be lower for electrodeposited (ED) nanocrystalline Ni. The origin of better corrosion resistance for nanocrystalline metals was a subject of discussion and was attributed to increased breakdown potential and a better resistance to anodic dissolution as a result of lower porosity. Evidence provided to substantiate these hypotheses or cogent reasons for not observing a higher localized corrosion even in the presence of such a high volume fraction of grain boundaries has not been conclusive. The discovery of grain boundaries in ED nanocrystalline Ni to be predominantly coherent low-sigma coincidence site lattices ($\Sigma 3$ CSLs) has been attributed to the possible origin of its superior corrosion resistance. Recently, it has been demonstrated that grain refinement, independent of the processing route, leads to the formation of high-angle boundaries and provides corrosion resistance to down-hole alloys. This breakthrough and recent advances in processing technology to engineer high-strength corrosion-resistant metallic nanomaterials will enable the design of nanostructured metallic materials specifically for the oil and gas industry. However, for the case of bulk nanostructured materials processed via SPD, it is necessary to note that up to now, there is no unified opinion on their corrosion properties. In case of nanostructured (Unalloyed Commercially Pure) CP Ti, it was shown that simultaneous enhancement of both mechanical strength and corrosion properties in aggressive environments, such as HCl and H_2SO_4 solutions, are possible. Corrosion resistance of SPD-processed Ti is determined not only by grain size, but also by the crystallographic texture developed during SPD processing. Particularly, basal planes of CP Ti offer higher corrosion resistance independently of grain size. Dislocation density can also significantly affect corrosion properties of this material. Very recently, it was demonstrated that corrosion resistance of the SPD-processed CP Ti can be further improved without any loss of mechanical strength via proper annealing treatments due to reduction of dislocation density and residual stresses. Very similar results were reported for SPD-processed Mg alloy AZ61 tested in 0.1 M NaCl solution [18] (Fig. 7.11).

7.4.7 Reliability of Nanostructured Materials

Reliability is the probability of survival in service of a component at a predetermined property level for a given length of time. Therefore, it is primarily concerned with the statistics of failure time, which may be regarded as the most important variable in Reliability Theory. It establishes an absolute limit to the lifetime of a component. For example, in the case of structural nanomaterials, parameters like strength and ductility have to be related to the failure time. Concepts associated with reliability are applied in all fields of engineering design and have been used extensively in the case of electronic and structural components and in materials assemblies that may comprise several dissimilar materials. It is self-evident that other considerations like materials selection, economics, productivity, strength-to-weight ratio and flexibility

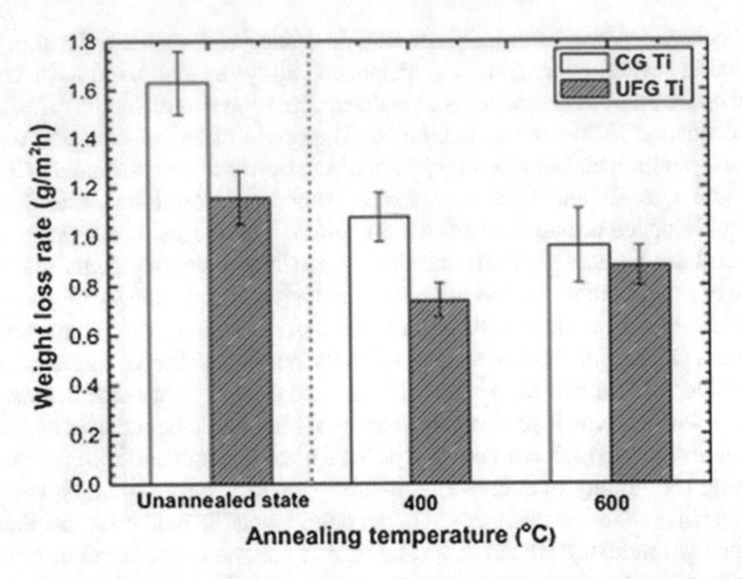

Fig. 7.11 Weight loss measurement test results of the CG Ti and UFG Ti in a 0.5 M H_2SO_4 solution before and after annealing at different temperatures for 2 h Reprinted with permission from © Corrosion Science, 89, 331–337 (2014)

become relevant only if the reliability of components in service can be ensured. Reliability of advanced materials is even more difficult to accomplish because they are used in the most demanding conditions, where conventional materials will prove to be inadequate. The ability of nanostructured materials to display properties that are different from those of their conventional analogues or are entirely new has made the currently available databases irrelevant. Therefore, reliability of nanostructured materials is primarily concerned with structure–property correlations with a view to ensuring the retention of the assumed design values of properties during the service life of a component and the capacity to predict its residual life at any given point in time. This suggests the need for a multidisciplinary, systems-oriented approach to producing nanostructures that function reliably in service. Common features that need to be addressed in order to increase the reliability of nanostructures are: surface quality, fracture, fatigue, creep, interface stability, phase stability and thermal stability [19].

7.4.8 Thermal Properties of Nanostructures

The second law of thermodynamics states that for large systems and over long times the entropy production rate is necessarily positive. However, the fluctuation theorem predicts measurable violations of the second law for small systems over short timescales. This has been shown experimentally by following the trajectory of a colloidal particle in an optical trap. An optical trap is formed when the micron sized particle with a refraction index higher than that of the surrounding medium is located within a focused laser beam; the refracted rays eventually exert a force on the particle which can be resolved to 2×10^{-15} N. From observing the particle's position and the optical forces acting on the particle, the entropy production rate Σ can be determined. It turns out that entropy consuming, i.e., second law-defying events, can be discerned for micron-sized particles on the timescales of seconds. This is particularly important to applications of nanomachines and molecular motors. As these nanomachines become smaller, the probability that they will run thermodynamically in reverse inescapably increases [20].

7.4.9 Thermal Conductance

In nanosystems the classical picture of a diffusive heat flow mechanism is often not applicable because the phonons or electrons that carry heat have mean free paths similar to or larger than the nanoscale feature size. This is a challenge for heat removal in microelectronic devices which already involve features with sizes of the order of the mean free path. The thermal conductance $\kappa(Vg)$ of electrons in a semiconductor quantum wire at low temperatures shows a quantized behavior in dependence of a gate voltage Vg. This originates from the plateaus in the electrical conductance $G(Vg)$ quantized in units of $G_O = 2e^2/h$.

7.4.9.1 Lattice Parameter

The lattice parameter of Pd nanoparticles in a polymer matrix decreases by about 3% when the particle size decreases from 10 to 1.4 nm. A similar behavior is observed for Ag nanoparticles.

7.4.9.2 Phase Transitions

Phase transitions in confined systems differ, as shown in the case of the melting transition above, from those of bulk materials and strongly depend on particle size, wetting, as well as the interaction of a nanoscale system with a matrix or a substrate. Phase transitions in nanosized systems have been investigated for superfluidity where the critical behavior of superfluid He in an aerogel deviates from that of bulk He. In superconductivity the superconducting transition temperatures of Ga or In nanoparticles in vycor glass are shifted to values higher than in the bulk materials. The Curie temperature T_C of ferromagnetic nanolayers decreases with decreasing layer thickness. Liquefaction in nanopores or order—disorder phase transitions near surfaces was also found to differ from that of bulk systems. Two examples of solid–solid phase transitions on the nanoscale will be sketched in the following. Under elevated pressures PbS undergoes a B1-to-orthorhombic structural phase transition. For PbS nanoparticles in a NaCl matrix, the transition is shifted to higher pressures when the particle size is reduced. The stability of crystal structures at ambient conditions depends on the size of ZrO_2 nanocrystals which exhibit a tetragonal structure for small sizes and the bulk orthorhombic structure for larger sizes. In this oxide, the lattice strain varies with the grain sizes giving rise to a variation of the Landau free energy so that for small grain sizes the tetragonal phase is stabilized whereas for grain sizes $d > 54$ nm the bulk orthorhombic phase appears [21].

References

1. Xu J, Li Y, Xiang Y et al (2018) A super energy mitigation nanostructure at high impact speed based on buckyball system. PLUS ONE 8:4697–4705
2. Whitesides GM, Kriebel JK, Mayers BT (2005) Self-assembly and nanostructured materials. In: Huck WTS (ed) Nanoscale assembly. Nanostructure science and technology. Springer, Boston, MA
3. Wang H, Li M, Li X et al (2015) Preparation and thermal stability of nickel nanowires via self-assembly process under magnetic field. Bull Mater Sci 38:1285–1289
4. Singh MR, Cox JD, Brzozowski M (2014) Photoluminescence and spontaneous emission enhancement in metamaterial nanostructures. J Phys D Appl Phys 47:085101–085107
5. Schroer MD, Petta JR (2010) Correlating the nanostructure and electronic properties of InAs nanowires. Nano Lett 10:1618–1622
6. Zhang X, Li X, Mara N et al (2010) Mechanical behavior of nanostructured materials. Metall Mater Trans A 41:777
7. Yang P (2003) The chemistry of nanostructured materials. World Scientific Publishing Co. Pte., Singapore
8. Huang JY, Liao XZ, Zhu YT et al (2003) Grain boundary structure of nanocrystalline Cu processed by cryomilling. J Philos Mag 83:1407–1419
9. Sabirov I, Enikeev NA, Murashkin MY et al (2015) Bulk nanostructured materials with multifunctional properties. Springer, Berlin. https://doi.org/10.1007/978-3-319-19599-5
10. Tsuji N (2007) Unique mechanical properties of nanostructured metals. J Nanosci Nanotechnol 7:3765–3770
11. Koch CC, Youssef KM, Scattergood RO et al (2005) Breakthroughs in optimization of mechanical properties of nanostructured metals and alloys. Adv Eng Mater 7:787–794
12. Pippan R, Hohenwarter A (2016) The importance of fracture toughness in ultrafine and nanocrystalline bulk materials. Mater Res Lett 4:127–136
13. Toutanji HA, Friel D, El-Korchi T et al (1995) Room temperature tensile and flexural strength of ceramics in AlN-SiC system. J Eur Ceram Soc 15:425–434
14. Tellkamp VL, Lavernia EJ, Melmed A (2001) Mechanical behavior and microstructure of a thermally stable bulk nanostructured Al alloy. Metall Mater Trans A 32:2335–2343
15. Bandyopadhyay AK (2008) Nanomaterials. New Age International (P) Ltd, New Delhi
16. Murashkin M, Yu I, Sabirov Sauvage X (2016) Nanostructured Al and Cu alloys with superior strength and electrical conductivity. J Mater Sci 51:33–49
17. Shi D, Aktas B, Pust L et al (2001) Nanostructured magnetic materials and their applications. Springer, Berlin, Heidelberg
18. Nykyforchyn H, Kyryliv V, Maksymiv O et al (2016) Formation of surface corrosion-resistant nanocrystalline structures on steel. Nanoscale Res Lett 11:51–56
19. Chang SJ, Meen TH, Prior SD (2014) Nanostructured materials for microelectronic applications. Adv Mater Sci Eng 383041

20. Bhushan B (2004) Nanotribology and nanomechanics. Springer handbook of Nanotechnology, p 1222
21. Schaefer HE (2010) Nanoscience. The science of the small in physics, engineering, chemistry, biology and medicine. Springer, Berlin Heidelberg, New York

Chapter 8
Nanostructured Materials—Design and Approach

Abstract Nanostructured material handling, properties and applications are widely important for promptly advancing field. This chapter focuses on important synthesis methods for production of nanostructural materials and selected properties of synthesized materials.

8.1 Synthesis of Nanostructured Materials

The nanostructural design of materials is schematically illustrated in Fig. 8.1 the scheme modifies and further develops a well-known concept of contemporary creation of novel materials through the integration of theory and modeling, structure characterization, processing and synthesis, as well as the properties of studies. In addition, nanostructuring of bulk materials deals with a far larger number of structural parameters related to the grain size and shape, lattice defects in the grain interior, as well as with the grain boundary structure, and also the presence of segregations and second-phase nanoparticles. This provides an opportunity to vary the transport mechanisms and therefore can drastically increase the properties. For example, nanostructuring of bulk materials by super plasmon diffusion (SPD) processing not only permits a considerable enhancement of many mechanical and physical properties but also contributes to the appearance of multifunctional materials. In this respect, one can anticipate that already in the near future, nanostructuring of materials by various processing and synthesis techniques may provide many new examples in the development of materials with superior properties for advanced structural and functional applications [1].

Two broad strategies are commonly employed for generating nanostructures. The first is bottom-up: that is, to use the techniques of molecular synthesis, colloid chemistry, polymer science, and related areas to make structures with nanometer dimensions. These nanostructures are formed in parallel and can sometimes be nearly identical, but usually have no long-range order when incorporated into extended materials. The second strategy is top-down: that is, to use the various methods of lithography to pattern materials. Currently, the maximum resolution of these patterns is significantly coarser than the dimensions of structures formed using bottom-up

T. D. Thangadurai et al., *Nanostructured Materials*, Engineering Materials,
https://doi.org/10.1007/978-3-030-26145-0_8

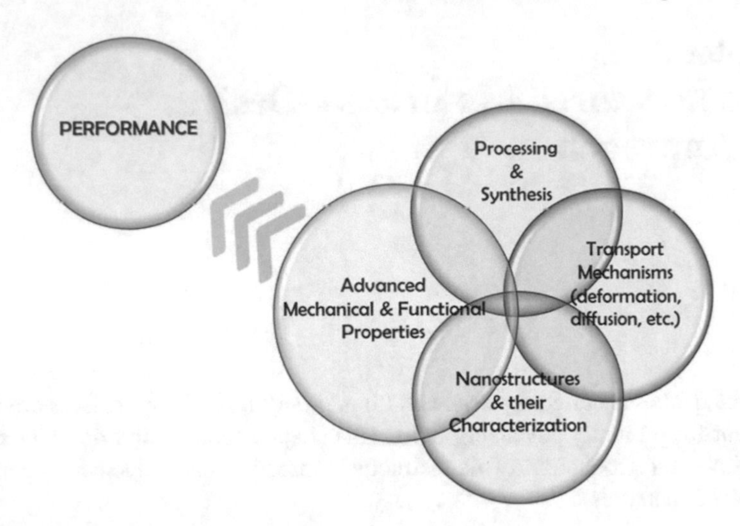

Fig. 8.1 Principles of nanostructural design of bulk nanostructured material

methods. Materials science needs an accessible strategy to bridge these two methods of formation, and to enable the fabrication of materials with the fine resolution of bottom-up methods and the longer-range and arbitrary structure of top-down processes. This bridging strategy is self-assembly: that is, to allow structures synthesized bottom-up to organize themselves into regular patterns or structures by using local forces to find the lowest-energy configuration, and to guide this self-assembly using templates fabricated top-down [2].

Quantum confinement has become a powerful tool for creating new structures with extraordinary properties. In creation, quantum-confined structures have been widely used in optoelectronic device technology. Then sensor applications based on quantum dots experiments existed in the field with the help of semiconductor nanocrystals. The possibility of having high quality, industrially scaled-up, biocompatible quantum dot nanocrystals has made a real breakthrough in the biological and medical fields.

Quantum dots significantly improve the sensing property in applications such as cellular assays, cancer detection, or DNA sequencing. There are different techniques for fabricating quantum dot structures depending on the material and also on the confinement strategy has two different categories; (i) QDs made by semiconductor growth techniques onto wafers and other, (ii) Synthesis of QD nanocrystal colloidal suspensions [3].

8.2 Nanostructure Synthesis and Fabrication Methods

There is various growth techniques employed for the fabrication of nanostructures, and to synthesis nanowires, nanobelts, nanocombs, nanotubes, and nanosquids.

1. Physical Vapor Deposition (PVD)
   a. Laser-Assisted Catalytic Growth
   b. Thermal Evaporation
   c. Radiofrequency Magnetron Sputtering.
2. Chemical Vapor Deposition (CVD)
   a. Thermal Chemical Vapor Deposition
   b. Metal–Organic Chemical Vapor Deposition (MOCVD).
3. Solution based synthesis methods
   a. Hydrothermal Synthesis
   b. Hydrolysis
   c. Aqueous Chemical Growth
   d. Electrospinning.

8.2.1 Physical Vapor Deposition

Physical vapor deposition (PVD) is vapor–solid (VS) deposition techniques that involve the generation of reactant vapors by physical processes such as heats, plasmas, and lasers.

8.2.1.1 Laser-Assisted Catalytic Growth

Laser-assisted catalytic growth (LCG) was among the earlier approaches for the growth of semiconducting nanowires. This technique involves the use of a pulsed laser to ablate a target that containing the element(s) desired in the nanowires and the metal catalyst component. In this technique the target was kept inside the growth chamber of a tube furnace. The vaporized compounds were then grown into nanowires of Si, GaN, InP, and GaP, through the vapor–liquid–solid (VLS) mechanism. This approach was used to prepare bulk quantities of single-crystalline nanowires with diameters of as small as 3 nm and lengths as long as 30 mm.

This technique was similar to the growth of single-wall carbon nanotubes by laser vaporization [4] (Fig. 8.2).

8.2.1.2 Thermal Evaporation

Thermal evaporation was one of the most widely used techniques for the growth of alternative nanostructures. In most cases, horizontal furnaces are used with a tubular reaction chamber constructed by an alumina tube or a quartz tube. This was the

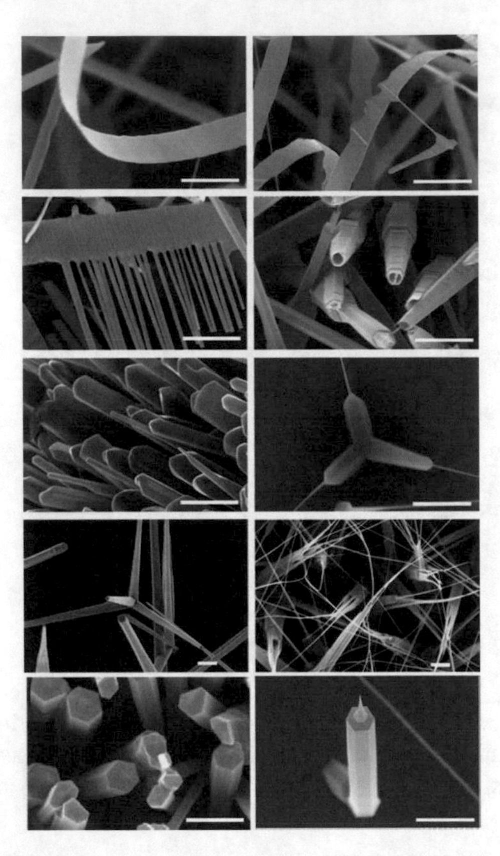

Fig. 8.2 ZnO nanostructures as nanobelts, nanopropellers, nanocombs, nanotubes, nanoswords, nanotripods, nanotetrapods, nanosquids, nanorods, and nanotips (left to right, top to bottom), scale bar 1 μm. Reprinted with permission from © Sensors Based on Nanostructured Materials, Springer, 59–78 (2009)

technique used for the growth of nanobelts of SnO_2, ZnO, In_2O_3, and CdO. For example, the growth of SnO_2 nanobelts was conducted in a horizontal tube furnace with an alumina tube chamber. SnO_2 powders were used as the source material that was evaporated at 1350 °C under a pressure of ~200–300 Torr and Ar gas flow rate of 50 sccm. These vapors were then condensed as belt-like nanostructures in a narrow region downstream where the temperature was ~900 to 950 °C. This vapor–solid condensation process did not involve the use of other metal catalysts, although Sn itself could have mediated the formation of SnO_2 nanobelts.

This approach was used for the growth of interesting diskettes of SnO_2 in a horizontal tube furnace as shown in Fig. 8.3. In this case SnO_2 powders were vaporized at 1050 °C under a pressure of 500–600 Torr with the flow of Ar gas. SnO_2 diskettes were deposited at the low temperature (LT) region (200–400 °C). By using the same approach, Ga_2O_3 nanoribbons and nanosheets were produced by thermal evaporation of GaN powders in a horizontal alumina tube furnace at 1100 °C. These Ga_2O_3

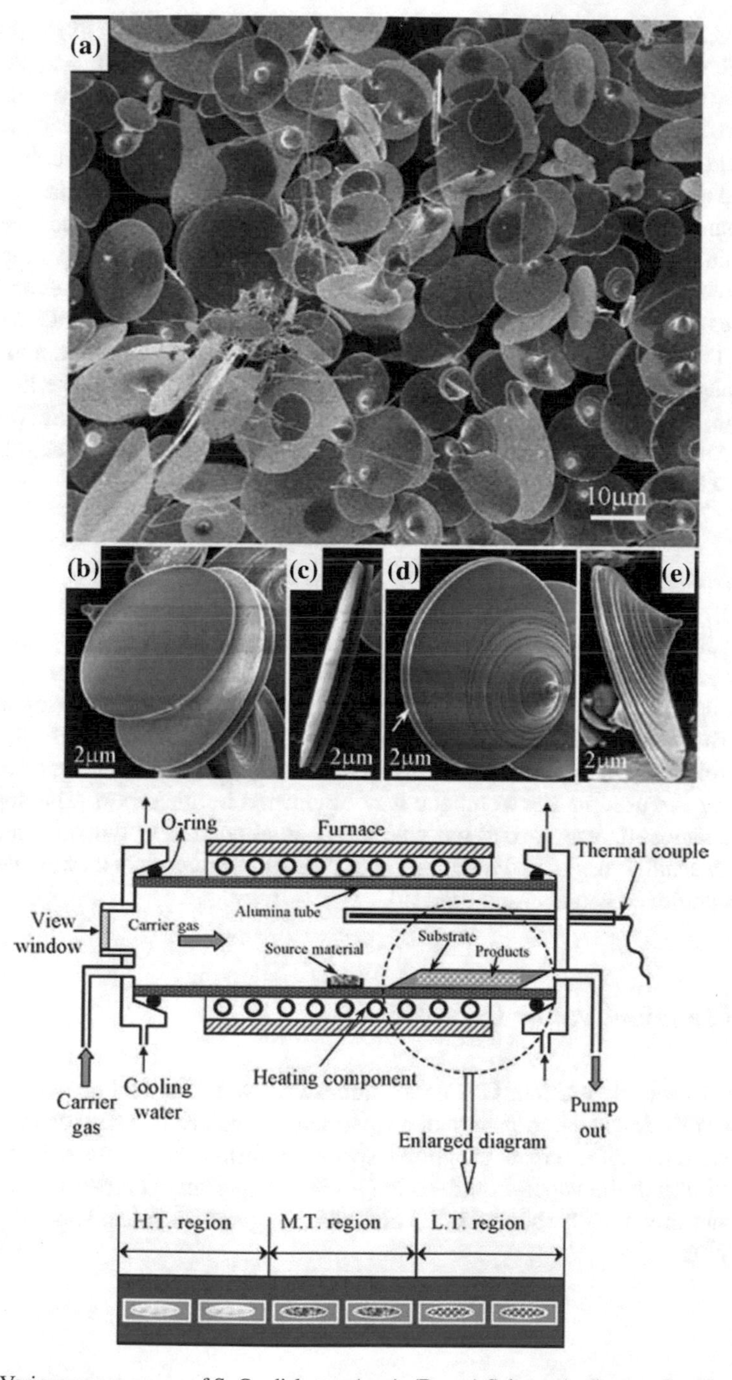

Fig. 8.3 Various appearances of SnO_2 diskettes (**a–e**). (Down) Schematic diagram for a horizontal tube furnace for the growth of various oxide nanostructures by thermal vaporization. Nanostructures can be condensed at high (H.T.), medium (M.T.), and low temperature (L.T.) regions downstream. Reprinted with permission from © Progress in Materials Science, 66, 112–255 (2014)

nanostructures were condensed at 800–850 °C under a pressure of 300 Torr created by the flow of Ar gas at a rate of 50 sccm. Various ZnO nanostructures can be produced by thermal evaporation of Zn powders. ZnO nanotetrapods were synthesized in a horizontal tube furnace with a fused-quartz tube chamber at 900 °C. A quartz plate held with a few spherical Zn pellets (~3 mm in diameter) was inserted into the tube and vaporized in air ambient. After heating for about 2 min, ZnO nanotetrapods were found on the surface of the quartz plate. These ZnO nanotetrapods were used for humidity sensors. On the other hand, ZnO nanobelts can be grown by evaporation of Zn powder at 600 °C. Results indicate that the control of gas flow rates and partial pressures of Ar, O_2, and Zn vapors are important for the growth of ZnO nanobelts. Finally, the growth of ZnO nanowires in quantity of several grams was demonstrated by heating the mixture of Zn powders and NaCl to 600–700 °C in the flow of Ar (25 sccm) and O_2 (20 sccm) gases that create a pressure of 2 Torr. About 70–80% of the Zn powders were converted to ZnO nanowires when NaCl was used, otherwise 5–10% without NaCl [5].

8.2.1.3 Radiofrequency Magnetron Sputtering

Vapor–solid growth of nanostructured materials can also be obtained by other physical vapor deposition technique such as radiofrequency (rf) magnetron sputtering. For example, ZnO nanobelts were grown by using rf magnetron sputtering of a ZnO target. These nanobelts were deposited on sapphire substrates without the use of catalyst at relatively low pressure of 40 mTorr and an rf power of 300 W for 60 min. The sputtering gas used in this technique is not indicated in this report. The deposition of these nanobelts was carried out without external heating to the substrates. This approach usually results in the formation of ZnO films but nanobelts were deposited when sapphire substrates were used [6].

8.3 Chemical Vapor Deposition

Chemical vapor deposition (CVD) techniques involved chemical decomposition or chemical reaction in the formation of nanostructured materials. These chemical reactions required to either create the growth species or form nanostructures by the use of catalyst through the vapor–liquid–solid (VLS) mechanism. The two types of CVD techniques are, namely thermal CVD and metal–organic chemical vapor deposition (MOCVD).

8.3.1 Thermal Chemical Vapor Deposition

Various ZnO nanostructures can be grown by carbothermal reduction of ZnO powders either with or without the use of catalyst such as Au. In this case, the precursor ZnO and graphite powders were mixed and loaded into a ceramic boat, which was placed in a small quartz reaction tube. This reaction tube was then inserted into a larger quartz tube of a horizontal tube furnace. This approach was sometimes referred as double-tube thermal CVD. Oxidized Si substrates (or other substrates) were usually placed downstream to the source materials at the lower temperature zones. In most cases, these precursor powders were combusted at 1100 °C so that ZnO powders will be reduced into Zn and ZnO_x vapors in the presence of graphite powders. These Zn and ZnO_x vapors were then condensed as ZnO nanostructures in O_2 gas at a pressure of several Torrs. ZnO appeared as nanowires, nanobelts, nanopropellers, nanocombs, nanotubes, nanoswords, nanotripods, nanotetrapods, nanosquids, nanorods, nanotips, etc. We found that nanorods can be grown without Au catalyst. These ZnO nanorods can be transformed into nanotubes and nanosquids with appropriate cooling during the growth. These nanorods can also be transformed into long nanowires by placing an Au-coated substrate beside the sampling substrates during the growth, a so-called side-catalyst approach.

This technique was applied for the growth of ZnO nanocombs by heating the tube furnace at 900 °C for 30 min. Argon (100 sccm) and oxygen (1 sccm) were used in this case without Au catalyst. Under this oxygen-deficient ambient, Zn vapors/droplets might be formed that functions as a catalyst for the growth of ZnO nanocombs. These nanocombs were tested as the biosensors for glucose detection. ZnO nanowires can also be grown by using ZnO and graphite powders with germanium (Ge) as the catalyst. The Ge catalysts were supplied in two ways. In one method, GeO_2 powders were reduced by the graphite powders together with the ZnO powders. In the other method Ge dots were patterned on the SiO_2-coated silicon wafer by photolithography. ZnO nanowires were condensed at the temperature zone of 500–650 °C with diameters and lengths of 50–400 nm and 50–200 mm, respectively. It is interesting to note that the Ge catalysts remain at the tips of these nanowires with diameters much larger than those of the nanowires. In addition to ZnO, other nanostructured metal oxides were grown by this vapor–liquid–solid approach. For example, SnO_2 nanowires were grown directly on the electrodes of sensors and tested for NO_2 gas sensing. The synthesis of these SnO_2 nanowires was carried out by thermal evaporation of Sn metal powders at low temperatures of 600–700 °C. This was performed in the flows of O_2 gas at a flow rate of 10 sccm and growth pressure of 1–10 Torr. Gold films of 0.2–0.3 nm were used as the catalyst for the growth of these nanowires. The diameters of the resulted SnO_2 nanowires were ranged from 50 to 100 nm, which was determined by the growth temperatures. The growth of silicon oxide (SiO_x) nanowires was demonstrated by a simple thermal CVD approach. In this case, GaN powders were used as the Ga source, i.e., the metal catalyst. The Si sources can be the Si wafers, SiO powders, or silane (SiH_4) gas. The GaN powders were placed at the location with temperature around 1150 °C, no matter what Si source

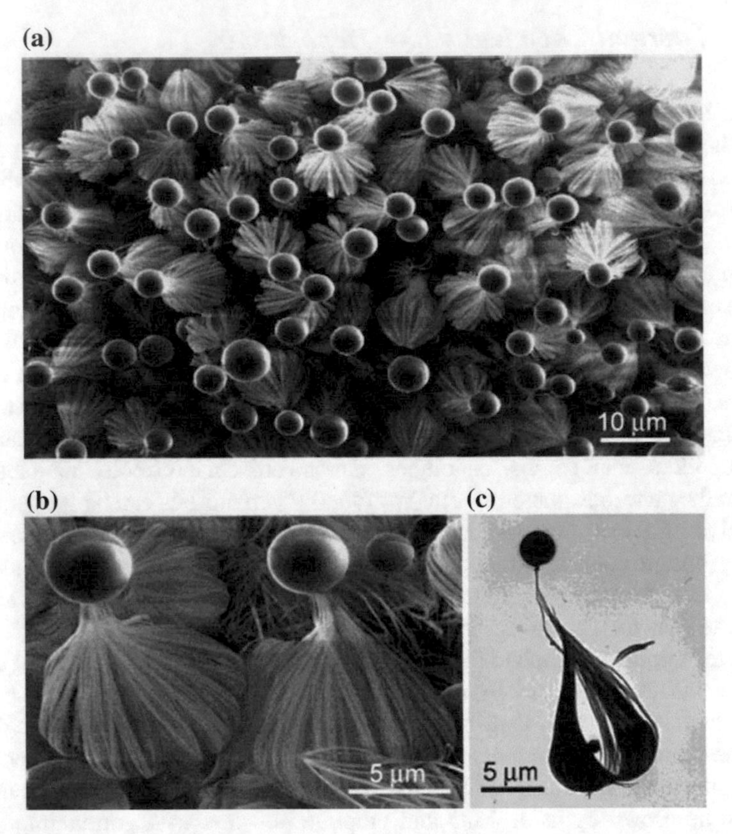

Fig. 8.4 Badminton-like SiO_x nanowire at different magnification (**a**, **b**), **c** TEM image of a badminton-like structure. Reprinted with permission from © Tsinghua Science and Technology, 10, 718–728 (2005)

was used. This technique was based on the VLS process. In fact, bundles of these SiO_x nanowires can be grown from a micrometer-sized Ga droplet. They can assemble into gourdlike, spindle-like, badminton-like, and octopus-like morphologies at different temperature zones. For example, the badminton-like structures can be formed at 980–1010 °C (Fig. 8.4a, b). These SiO_x nanowires exhibit branching growth features with a height of ~10 to 15 mm (Fig. 8.4c).

Thermal CVD was used for the growth of doped nanostructures. For example, Sb-doped SnO_2 nanowires were grown for gas sensor application. In this case, metal Sn and Sb powders mixed in the weight ratio of 10:1 were used as the evaporation sources in a thermal CVD system. Si substrates coated with 5 nm gold were placed downstream in the furnace.

The temperature at the center of quartz tube was 900 °C, and a constant flow of 1% oxygen and 99% nitrogen was maintained at a flow rate of 5 l/min. The as-grown nanowires have a diameter of 40–100 nm and lengths up to several tens of micrometers. The CVD has been a very useful and versatile technique. Nanostructures that were initially grown by other techniques have later been successfully grown by CVD. For example, instead of laser-assisted catalytic growth, silicon nanowires are now commonly grown by thermal CVD. This was carried out using Au nanoparticles as the catalyst and Silane (SiH_4, 10% in He) gas as the Si source [7].

8.3.2 Metal–Organic Chemical Vapor Deposition (MOCVD)

Thermal CVD techniques discussed so far are not always the suitable approach for the growth of certain nanostructures. For example, GaN nanowires have been grown by laser ablation and thermal CVD. High temperatures were needed to generate Ga vapor source in the thermal CVD technique, and Ni, Fe, and Au were used as the catalysts. The use of a solid Ga is technically simple but often leads to nonconstant vapor pressures for continuous growth of GaN nanowires. Metal–organic chemical vapor deposition (MOCVD) was found to be a more versatile technique for the growth of GaN nanowires. The MOCVD technique seems to overcome the vapor pressure problem and yield high-quality GaN nanowires In the MOCVD approach, trimethylgallium (TMG) and ammonia gas were used as Ga and N precursors. Silicon substrates or c-plane and a-plane sapphire substrates coated with 2–10 nm thin film of Ni, Fe, or Au were used for the growth at 800–1000 °C. The reaction was carried out in an oxygen-free environment at atmospheric pressure. TMG was kept cool in a—10 °C temperature bath. Nitrogen was used as a carrier gas and percolated through the TMG precursor and coupled with a second nitrogen line to give a total nitrogen flow rate of 250 sccm. Hydrogen and ammonia gases were supplied at a total flow rate of 155 sccm. The as-grown GaN nanowires were having diameters of 15–100 nm and lengths of 1–5 mm, oriented predominantly along the [210] or [110] direction. MOCVD technique was also used for growing nanostructured IrO_2 crystals. The nanostructured crystals were grown on a gold-coated quartz substrate and their gas-sensing properties were studied by quartz crystal microbalance (QCM) technique. The growth of these nanostructures was performed by using (methylcyclopentadienyl) (1,5-cyclooctadiene) iridium, (MeCp)Ir(COD), as the precursor. The quartz substrate temperature (Ts) was kept between 350 and 500 °C. High-purity oxygen was used as carrier gas at a flow rate of 100 sccm, which leads to a growth pressure of 7.2 Torr. The temperature of the precursor reservoir (Tp) was varied between 95 and 1058C. Different morphologies, such as nanoblades, layered columns, incomplete nanotubes, and square nanorods, were observed at various combinations of Ts and Tp [8].

8.4 Solution-Based Chemistry

In addition to the vapor phase growth techniques discussed so far, several solution phase deposition techniques have been used for the growth of various alternative nanostructures.

8.4.1 *Hydrothermal Synthesis*

Tungsten oxide nanowires ($WO_{2.72}$) were grown by hydrothermal technique and were tested for hydrocarbon sensing. Sensors based on $WO_{2.72}$ nanowires show high sensitivity for 50–2000 ppm of LPG (propane–butane mixture) at 200 °C as well as relatively short recovery and response times. These $WO_{2.72}$ nanowires were prepared by solvothermal synthesis. Tungsten hexachloride (1 g) was loaded into a 25 ml autoclave filled with ethanol up to 90% of its volume. Hydrothermal synthesis was carried out at 200 °C for 24 h. The product obtained by centrifugation was washed with ethanol. The as-grown nanowires were having diameters of 5–15 nm and lengths of 100–200 nm. ZnO nanorods can also be grown by hydrothermal techniques. Zinc and cetyltrimethylammonium bromide (CTAB) were used as the precursors. CTAB (1.5 g) was dissolved in deionized water (35 ml) to form a transparent solution. Then, zinc powders (1.8 g) were added to the above solution under continuous stirring. The resulting suspension was transferred into a teflonlined stainless steel autoclave (volume 40 ml) and sealed tightly. Hydrothermal treatments were carried out at 182 °C for 24 h. The autoclave was then cooled down. Next precipitates were collected, washed with deionized water for several times, and dried in air. The lengths of the ZnO nanorods are usually shorter than 1 mm and their diameters ranging from 40 to 80 nm. These nanorods were used for the sensing of various vapors including alcohol, LPG, gasoline, ammonia, etc. Vanadium oxide (V_2O_5) nanobelts coated with Fe_2O_3, TiO_2, and SnO_2 nanoparticles have been prepared by mild hydrothermal reaction. For the growth of the V_2O_5 nanobelts, nitric acid was added dropwise to a 0.1 M ammonium metavanadate solution until the final pH value of the solution reached about 2–3 under stirring. Solution obtained was transferred to a Teflon-lined autoclave and filled with deionized water up to 80% of the total volume. Then the autoclave was kept at 180 °C for 24 h. The final product was washed with deionized water and pure alcohol several times to remove any possible remnants. The as-grown nanobelts were tens of micrometers long with smooth surfaces, typically 60–100 nm wide and 10–20 nm thick. These nanobelts were then coated with the oxide nanoparticles for the sensing of alcohol, benzene, cyclohexane, gasoline, ammonia, etc. $NiFe_2O_4$ nanospheres, nanocubes, and nanorods were prepared by a hydrothermal method. For synthesizing these $NiFe_2O_4$ nanostructures, $Ni(NO_3)_2 \cdot 6H_2O$ and $Fe(NO_3)_3 \cdot 9H_2O$ were dissolved in deionized water to form a mixed solution with $[Na^{2+}] = 0.10$ mol/l and $[Fe^{3+}] =$ 0.20 mol/l. NaOH solution (6.0 mol/l) was added dropwise under stirring into 20.0 ml of the mixed solution until the desired pH value was attained to form an admixture.

In the next step, the admixture was transferred to a Teflon autoclave (50 ml volume) with a stainless steel shell up to 80% of the total volume. The autoclave was kept at 120–200 °C for 24–96 h and then cooled down. The final product was washed with deionized water and alcohol for several times and then dried. The length and diameter of the nanorods were about 1 mm and 30 nm, respectively; the side length of the nanocubes was about 60–100 nm. It was found that sensors based on $NiFe_2O_4$ nanorods were relatively sensitive and selective to triethylamine [9].

8.4.2 Hydrolysis

The growth of Fe_2O_3/ZnO core/shell nanorods was reported by a solution phase-controlled hydrolysis process of Zn^{2+} ions in the presence of Fe_2O_3 nanorods. One hundred milligram of Fe_2O_3 nanorods (~100 nm long, 25 nm in diameter) were dispersed in 200 ml deionized water by ultrasonication. Then 20 mg of zinc acetate ($Zn(Ac)_2 \cdot 6H_2O$)) was introduced to the solution, and the suspension was heated in an oil bath at 40 °C under vigorous stirring. Twenty milliliters of 5% ammonia was then added into the suspension for 30 min, and the reaction was maintained at the temperature for 1 h. The colloidal suspension was then centrifuged and sintered at 400 °C for 2 h to obtain Fe_2O_3/ZnO nanorods. These core/shell structures have diameters of ~30 nm.

8.4.3 Aqueous Chemical Growth

ZnO nanostructures were deposited on glass substrates by the aqueous chemical growth technique at ~95 °C. The growth process involves the use of an equimolar (0.01 M) aqueous solution of $Zn(NO_3)_2 \cdot 6H_2O$ and $C_6H_{12}N_4$ as the precursors. The solution and the substrates were then placed in glass bottles and heated at 95 °C for 1, 5, 10, and 20 h. After each induction time, the substrates were thoroughly washed with deionized water and dried in air. Flower-like aggregations of ZnO nanorods were deposited in all cases. The diameters of these nanorods are relatively big (~500 to 1000 nm) and increased with the deposition duration. Their lengths are usually <10 mm [10].

References

1. Whitesides GM, Kriebel JK, Mayers BT (2005) Self-assembly and nanostructured materials. In: Nanoscale assembly. Nanostructure science and technology. Springer, Boston, MA
2. Fiorani D, Sberveglieri G (1994) Fundamental properties of nanostructured materials. World Scientific Publishing Co. Pte. Ltd, Italy
3. Arregui FJ (2009) Sensors based on nanostructured materials. Springer, Spain

4. Horprathum M, Eiamchai P, Kaewkhao J et al (2014) Fabrication of nanostructure by physical vapor deposition with glancing angle deposition technique and its applications. AIP Conf Proc 7:1617–1624
5. Khanh LD, Binh NT, Binh LTT et al (2008) SnO_2 nanostructures synthesized by using a thermal evaporation method. J Korean Phys Soc 52:1689–1692
6. Hernandez MA, Alvaro R, Serrano S et al (2011) Catalytic growth of ZnO nanostructures by r.f. magnetron sputtering. Nanoscale Res Lett 6:437–444
7. Shariffudin SS, Herman SH, Rusop M (2014) Self-catalyzed thermal chemical vapor deposited ZnO nanotetrapods. Adv Mater Res 832:670–674
8. Fu K (2009) Growth dynamics of semiconductor nanostructures by MOCVD. Universitetsservice US AB, Stockholm, Sweden
9. Xu X, Wu M, Asoro M et al (2012) One-step hydrothermal synthesis of comb-like ZnO nanostructures. Cryst Growth Des 12:4829–4833
10. Nair SS, Forsythe J, Winther-Jensen B (2015) Directing the growth of ZnO nano structures on flexible substrates using low temperature aqueous synthesis. RSC Adv 5:90881–90887

Chapter 9
Functionalization of Nanostructures

Abstract Presently, coating on nanostructural materials is preferred for exact release of biological entities into human system for varies desired application like drug delivery system, and thus the functionalization is an important concept for nanostructural material development. The functionalization is made by coating the material surface, tuning the material size etc.

9.1 Aspects of Nanostructure System

Nanotechnology allows for a unique control of the material world, and any material comprising an average phase or grain size in the order of a nanometer (10^{-9} m) is defined as a nanostructured material. At this scale, man-made objects are able to gain access to cells as nanostructured materials have enhanced mechanical properties compared to conventional materials due to their ultrafine microstructure and ability to interface with living cells. Nanostructured materials of varying compositions play important roles in cancer therapeutics due to their passive tumor targeting property. In this context, inorganic metal complexes occupy a pioneer niche as they offer a multipurpose platform for drug design and development, and the nanostructures used in the functionalization of metallodrugs act as transport vehicles to deliver the metallodrugs to the biological target. Nanostructured systems, such as macromolecular systems, carbon nanotubes, dendrimers, metallacages, ceramic materials, liposomes, and lipid nanocapsules, as well as metal and polymeric nanoparticles are commonly used for drug delivery. In addition, the optimum loading and sustained release of metallodrugs from nanostructures are essential requirements for every drug delivery system. Therefore, it is important to ensure that metallodrugs are attached to the surface of nanoparticles with adequate strength to withstand the wear and tear during the passage to the target site. However, it is also crucial to ensure that the bond is weak enough to release the metallodrugs at the target site [1].

Nanostructured surfaces can be produced on conventional biomaterials by surface modification. Depending on the ways surfaces are modified, nano-functionalization

T. D. Thangadurai et al., *Nanostructured Materials*, Engineering Materials,
https://doi.org/10.1007/978-3-030-26145-0_9

techniques can be categorized into two groups, namely nano coating and film deposition as well as in situ surface nano-functionalization. These two types of techniques are often combined to produce surfaces with hybrid nanostructures such as a coating and a nanostructured zone.

Typical coating and film deposition techniques are plasma spraying, plasma immersion ion implantation & deposition (PIII&D), sol–gel, chemical vapor deposition (CVD), physical vapor deposition (PVD), cold spraying, self-assembly, and so on, whereas in situ surface modification techniques include laser etching, shot blasting, acid and alkali treatments, anodic oxidation, micro-arc oxidation, ion implantation, etc. The nano-functionalized surfaces also possess promising biological properties and therefore, the biological properties and clinical applications of titanium, titanium alloys, and related materials can be improved and widened by producing a nanostructured surface [2].

9.2 Chemistry of Nanostructure Functionalization

Nanostructuration may produce surfaces with controlled topography and chemistry that could help developing of novel implant surfaces with predictable tissue-integrative properties. It was indicated that the implant surface nanostructuration is a decisive factor in surface cell adhesion and growth, by modulating cell behavior and protein adsorption. Surface chemical composition and local atomic charges are also critical for protein adsorption and cell attachment. Recently, Calcium phosphate compounds CaPs have been widely used as coatings for metallic prostheses to improve their biological properties. Once the integration of titanium implants coated with biomimetic calcium phosphate has been investigated in pre-clinical comparative models. Studies have demonstrated a higher bone-to-implant contact for biomimetic calcium phosphate coatings than for uncoated ones. A common method of adding functionality to silica nanoparticles is by installing amino groups onto the surface. Amination is an ideal method to install functionality because amino groups react specifically and spontaneously with other functional groups. Installing amino groups on the surface of the particles, followed by conjugation of water-soluble molecules, increases biocompatibility. Silica nanoparticles themselves are biocompatible, but certain molecules such as PEG can enhance biocompatibility, stability, cell penetration, and water solubility [3].

9.3 Need for Functionalization

Functionalization has been used to conjugate drug molecules, polymers and organic groups to NPs. In addition, functionalization has also been shown to protect NPs against agglomeration and render them compatible in other phases. Functionalization also improves the physical, chemical and mechanical properties of NPs, which are synergetic. Functionalization of the NPs can be defined as the addition of a chemical functional group on their surface in order to achieve surface modification that enables their self-organization and renders them compatible. NPs have mainly been functionalized with thiols, disulfides, amines, nitriles, carboxylic acids, phosphine and biomolecules. The main goal of functionalizing NPs is to cover their surface with a molecule that possesses the appropriate chemical functionality for the desired application. In all cases, functionalization of the particles produces a drastic change in their surface properties. The surface chemistry of NPs is already an important aspect of their synthesis, since this property can be exploited to control their size and self-organization during formation. This can be achieved by complexing groups that bind on the surfaces during their formation and complex formation should not promote agglomeration [4].

9.4 Methods of Functionalization

There are two strategies for introducing functional groups to surfaces and NPs. The first method is direct functionalization, where the whole functional ligand is a bi-functional organic compound. In this approach, one of the functionally reactive groups is used to attach to the NPs surface and the second group contains the required active functionality. Direct functionalization is preferred because it only requires a single conjugation step. One limitation of the direct functionalization method is the incompatibility of the functional group F with the preparation process, for instance the modifying group may react with the particle surface.

The second method is post-functionalization, which is generally preferred because this strategy is more versatile and the nature of the functionalizing moieties may not be fully compatible with good control over the size and dispersion state of the particles in the solvent used for their synthesis. This describes a bifunctional compound where a binding chelating group is reacted first and the group of coupling site can be converted, in a second step, to the final functional group F. Post-functionalization of the particles requires a molecule to be grafted on the surface such that it has a structure that can be described as N-C-F (Nanoparticle-Chelating agent-Functional group). For post-functionalization, silane like compounds has been generally used. Hydrolysis reactions lead to the formation of a silane like coating around the particles with a fraction of the functional group orientated towards the outside of the particle. This process has been used in many systems, mostly in oxide systems, but it has also been investigated in the case of chalcogenides [5] (Fig. 9.1).

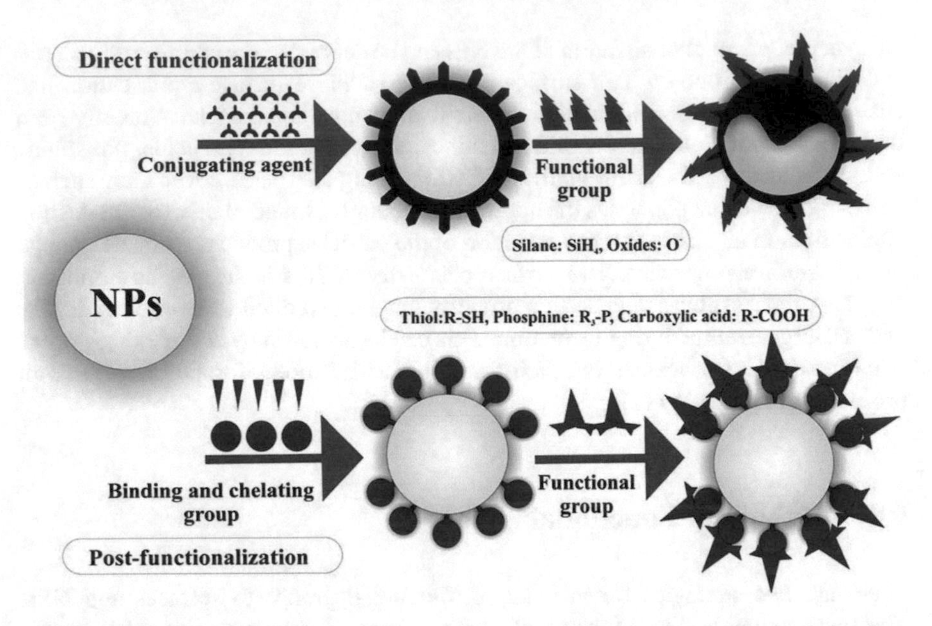

Fig. 9.1 Schematic representations of functionalization methods. Reprinted with permission from © Current Medicinal Chemistry, 17, 4559–4577 (2010)

9.5 Class of Functionalization

9.5.1 Thiol/Aminothiol

Thiol chemistry is one of the most developed functionalization methods used for the fabrication of functional NPs. Figure 9.2 describes the chemical functionalization reaction of thiols. Sulfur compounds naturally form strong coordination bonds with many metals including Ag, Cu, Pt, Hg, Fe, Si, and Au. Sulfur and organosulfur have a high affinity for metal surfaces and thus will adsorb spontaneously. Thiol or disulfide capped NPs can be prepared by either the direct functionalization method where the metal precursor and the protective ligand are reacted simultaneously. Another method based on thiol chemistry is the post functionalization method, where the sulfur compounds are grafted on the surface of presynthesized NPs that are covered by solvent molecules and are thus replaced by sulfur containing ligands [6].

9.5.2 Bio-functionalization

NPs functionalized with biomolecules have recently attracted great interest because the resulting hybrid materials have proven to be useful as drug carriers. Small biological molecules such as amino acids can be anchored to the surface of the NPs,

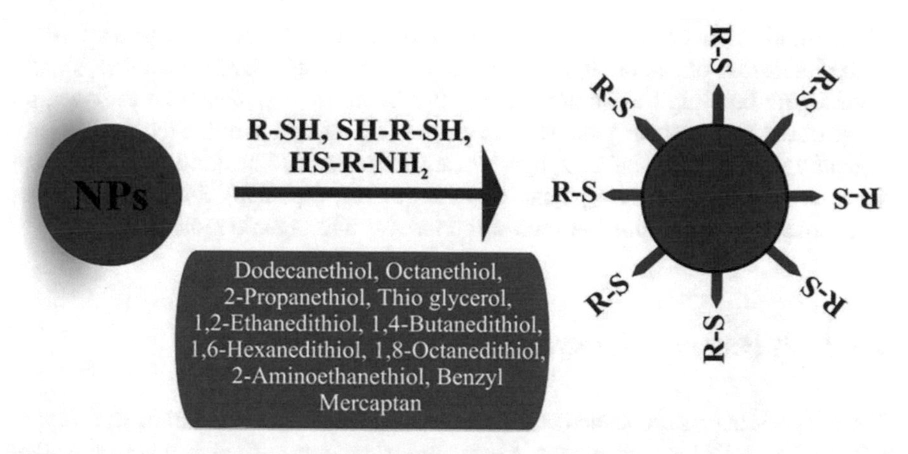

Fig. 9.2 Schematic of the thiol group based functionalization. Reprinted with permission from © Current Medicinal Chemistry, 17, 4559–4577 (2010)

opening the possibility of attaching functional biomolecules to bio FNPs. Au NPs functionalized with proteins and investigated the mechanical properties of the resulting NPs, which could be used to tune the properties of biomolecular films. The use of N-(2-aminoethyl)-APS (AEAPS) for the surface modification of iron oxide NPs have become a popular method for the covalent attachment of antibodies such as anti-CD34. NPs modified with molecules such as organosilane, vinyl alcohol/vinyl amine co-polymer were used for subsequent conjugation of antibodies, which could then be utilized for targeting specific biological cells and tissues. molecular targeting camptothecin (anticancer) drug delivery system using BH3 and LHRH peptides functionalized polymeric NPs to treat ovarian cancer [7].

9.5.3 Asymmetric Group

Asymmetric nanostructures are more versatile building blocks compared to their symmetric counterparts. For example, Au NPs were asymmetrically modified with single strand DNA, which can be used as the building block to prepare more complex structures such as dimers and trimmers, also asymmetric diblock Au-polymer nanorods can self-assemble into bundles, tubes, and sheets Similarly, Au shell structures were grown on a silica surface to produce Au cups or caps. The asymmetric structure in the nano rod system has also been realized using the chemical vapor deposition technique, which is known as the direct functionalization method. Likewise, asymmetric rods can be synthesized using template directed growth. However, when this approach is used, relatively large particles are produced. For example, the Au-metal oxide asymmetric dimers can only be prepared from metal oxide particles larger than 200 nm. The Au-polymer asymmetric nanorods have diameters around

200 nm and lengths over 1 μm. As a result, they can only be used to prepare particles with dimensions of μm or larger. To synthesize nanometer sized assemblies, smaller asymmetric building blocks are needed. Furthermore, it is desirable to develop a range of methods that can yield such structures. The structure of the tetrapod is topologically similar to that of a sp^3 hybridized carbon atom. This kind of structure can serve as a building block to prepare superstructures, especially *3D* superstructures, by mimicking the bonding between carbon atoms and organic molecules [8].

9.5.4 Polymers in Functionalization

The polymer matrix embedded in the NPs plays a major role in dictating the compatibility of the NPs for applications in harsh environments such as in acidic and alkali solutions. Also polymer NPs are widely used in pharmaceuticals application particularly cancer therapy and drug targeting. In polymeric nanocomposites synthesis, NPs or nanofibers have been used as fillers to improve the properties of the nanocomposites such as strength, which is very important in tissue engineering applications [9].

9.5.5 Functionalization of Metals

The main purpose of surface functionalization of pure Mg or Fe and alloys is to control their degradation rate. In the particular case of Mg, the degradation after surface functionalization can be highly mitigated but not stopped. Contrary, when considering Fe, the surface functionalization aims an increased corrosion rate. The underneath substrate degradation process will be constrained by the presence of cracks or defects on the coating. Therefore, the durability and integrity of the coating is crucial in biomedical applications. An improved surface biocompatibility and/or drug vehicle for controlled release can be regarded as a plus for pure Mg, Fe or Zn and alloys, although in the specific case of Zn, the surface functionalization of Zn should induce no more than slight changes in the corrosion rate of this metallic biomaterial [10].

9.5.6 Rare-Earth in Functionalization

The design of nanostructured materials containing (Rare Earth) RE elements, either as major components or as dopants, has paved the way for the development of new applications. In particular, nanoparticles (NPs) possess the size (ranging from 1 to

100 nm) in which most of the biomolecular interactions take place, so the incorporation of RE into NPs allows their use in many different biomedical applications, including bioimaging, biosensing, targeting, drug delivery and other therapies.

9.5.6.1 Functionalization of RE-Based Nanostructures

Solvothermal-based methods are those carried out in closed vessels under autogenous pressure above the boiling point of the solvent. They have been applied for the synthesis of many different nanomaterials. RE chlorides and nitrates are normally used as precursors. When using water as solvent (hydrothermal synthesis), in some cases, no additives are required to control particle size and shape, as in the synthesis of $NaREF_4$ (RE = Pr, Sm, Gd, Tb, Dy, Er) nanotubes, which were obtained using a two-step procedure. Firstly, $RE(OH)_3$ nanotubes of dimensions of 300 × 50 nm were synthesized after submitting at 120 °C a aqueous solution of the corresponding RE nitrates at pH 14. Such $RE(OH)_3$ nanotubes were then treated at 120 °C with aqueous solutions containing HF and NaF to obtain $NaREF_4$ nanotubes. $CePO_4$ nanotubes were also prepared without additives after hydrothermal reaction of $Ce(NO_3)_3{\cdot}6H_2O$ and $(NH_4)_2HPO_4$ solutions at pH 1 and 180 °C. Likewise, $REPO_4$ (RE = La, Ce, Pr, Nd, Sm, Eu, Gd, Tb, Dy, Ho, Er, Tm, Yb, Lu, Y) nanotubes with diameters ranging from 5 to 120 nm were obtained after treatment at 150 °C of aqueous basic solutions containing the corresponding RE nitrates and $NH_4H_2PO_4$. Finally, $CePO_4{:}Tb^{3+}$ and $LaPO_4{:}Eu^{3+}$ nanorods were also synthesized in a water/ethanol mixture at 120 °C from the lanthanide chloride and H_3PO_4. In all these examples, an optimization of the reaction pH was carried out. In most hydrothermal syntheses, however, hydrophilic additives are needed to produce uniform NPs. Among them, the most common is sodium citrate, which is able to form strong complexes with the REs, as previously commented. Several examples can be cited in which these complexes have been used in aqueous solutions such as the synthesis of $NaGdF_4$ nanorods (100 nm) at 180 °C from solutions containing $Gd(NO_3)_3$ and NaF, and the synthesis of $BaGdF_5$ porous nanospheres of sizes ranging from 65 to 119 nm from solutions containing gadolinium and barium nitrate and $NaBF_4$ at 180 °C in butyl amine, which was used to increase the pH. Some RE phosphates have been also synthesized in this way as it is the case of $YPO_4{:}Ln^{3+}$ (Ln = Tb, Eu, Dy) nanorods (60 × 35 nm), synthesized at 180 °C from acid aqueous solutions of the corresponding RE chloride, sodium citrate and $Na_5P_3O_{10}$, and $YPO4{:}Ln^{3+}$ (Ln = Tb, Eu) spherical-like NPs obtained after hydrothermal treatment at 180 °C of basic aqueous solutions of $Y(NO_3)_3$, sodium citrate and Na_3PO_4.

Other hydrophilic additives used for the hydrothermal synthesis of RE-based NPs are PEG, Ethylenediaminetetraacetic acid (EDTA), PEI, PAA and polyvinylpyrrolidon (PVP). PEG has been used, for example, for the hydrothermal synthesis at 190 °C of $BaLuF_5$ NPs (24 nm) from solutions containing PEG, $BaCl_2$, $Lu(NO_3)_3$ and NH_4F. Likewise, $BaGdF_5$ nanospheres have been synthesized from aqueous basic solutions containing EDTA $BaCl_2$, $GdCl_3$ and $NaBF_4$ at 180 °C. PEI has been used for the solvothermal synthesis in EG at 200 °C of $NaGdF_4{:}Yb^{3+}$,

Ln^{3+} (Ln = Er, Tm, Ho) cubic NPs (40 nm) from the corresponding RE chloride, NH_4F and PEI; $KGdF_4$:Ln^{3+} (Ln = Gd, Eu, Tb, Dy) roughly spherical NPs (25 nm) were obtained from KCl, RE chloride and NH_4F solutions in EG at 200 °C. Eu^{3+} doped calcium apatite 191 × 40 nm nanospindles have been synthesized in water at 180 °C from calcium nitrate, sodium phosphate monobasic and PAA at basic pH. Finally, PVP has been used for the synthesis of 50 × 10 nm $GdPO_4$:Ln^{3+} (Ln = Ce + Tb, Eu, Yb + Er) nanorods by treatment at 180 °C of aqueous solutions of $Gd(NO_3)_3$ and Na_2HPO_4. Organic hydrophobic additives such as linoleic or oleic acid, often in presence of sodium linoleate or oleate, are also frequently used for the solvothermal synthesis of highly uniform RE-based NPs in organic media, mostly in alcohol and in alcohol/water solutions. Typical examples are the synthesis of highly uniform REF_3 and $NaREF_4$ fluorides from the corresponding RE acetate and NaF in water/ethanol-containing linoleic acid or sodium linoleate solutions at temperatures ranging from 100 to 200 °C. This procedure yields rounded NPs of $NaYF_4$, YbF_3 and LaF_3 of around 4–12 nm diameter and rice-like YF_3 NPs of 500 × 100 nm. Other RE fluoride systems synthesized following this strategy are LaF_3:Ln^{3+} (Ln = Ce/Tb and Yb/Er) and $NaLaF_4$ nanorods, obtained from the corresponding RE nitrate, NaOH, NaF and oleic acid in a water/ethanol solution treated at 190 °C and $NaYb_{1-x}Gd_xF_4$:Ln^{3+} (Ln = Eu^{3+}, Tb^{3+}, Ce^{3+}/Eu^{3+}, Ce^{3+}/Tb^{3+}, Ce^{3+}/Eu^{3+}/Tb^{3+}, Ho^{3+}, Er^{3+}, Tm^{3+} and Er^{3+}/Tm^{3+}) NPs obtained from the corresponding RE nitrate, NaOH and NH_4F at temperatures ranging from 130 to 230 °C. In the last case the Gd content played a key role in the final morphology of the NPs, which ranged from $NaYbF_4$ nanocubes to $NaYb_{1-x}Gd_xF4$ nanorods.

Round-shaped NPs for REF_3 (RE = La to Eu) and rice-like NPs for REF_3 (RE = Gd to Yb) have also been obtained at temperatures ranging from 100 to 200 °C from solutions of RE nitrates in water and ethanol with sodium linoleate and linoleic acid as additives. Some amines have also been employed instead of fatty acid derivatives. For example, GdF_3 nanowires (length $\leq$ 40 nm) were synthesized at 240 °C from a solution of $Gd(NO_3)_3{\cdot}xH_2O$ and HF in octylamine/ethanol, and rice-like YF_3 and YbF_3 NPs have been obtained from ethanol/water solutions of RE nitrates, linoleate acid, octadecylamine and NH_4HF_2 treated at 100–200 °C. Other RE-based phases have also been synthesized following this method. For example, YPO_4:Eu^{3+} hexagonal-like NPs of around 16 nm were obtained from the RE nitrates, NaH_2PO_4, NaOH and oleic acid in a water/ethanol solution at temperatures ranging from 100 to 140 °C. Likewise, $REPO_4{\cdot}H_2O$ (RE = Y, La-Nd, Sm–Lu) of different morphologies (sphere-like, rod-like and polygon-like, with sizes ranging from 8 to 60 nm, depending on the RE) were synthesized from RE nitrate, NaH_2PO_4, oleic acid, and NaOH solutions in water and ethanol treated at 140 °C. However, the use of hydrophobic additives finally produces hydrophobic NPs, thus demanding further functionalization treatments. The main advantages of hydrothermal methods with or without hydrophilic additives over the solvothermal method using hydrophobic additives are the final hydrophilic character of the NPs, their high level of crystallinity and the fact that in contrast to most organic solvents water is not toxic. However, these methods also present some drawbacks such as the larger particle size that usually results in comparison with that produced by the organic solvent-based methods [11] (Fig. 9.3).

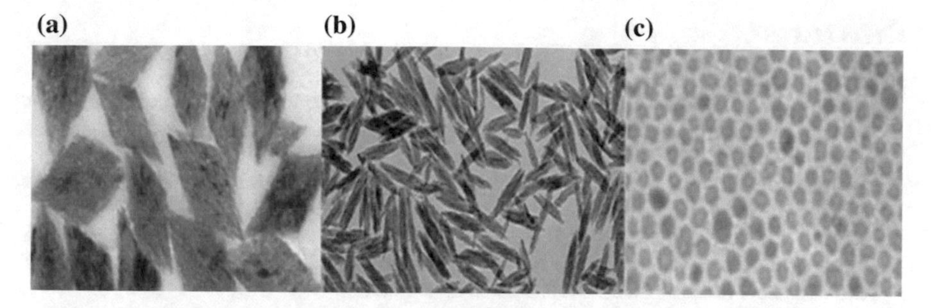

Fig. 9.3 (**a**) YF_3 NPs synthesized from yttrium acetylacetonate and the IL [BMIM] BF_4 in ethylene glycol at 120 °C. (**b**) $Ca_5(PO_4)_3)(OH):Eu^{3+}$ nanospindles synthesized through a microwave-assisted hydrothermal method at 180 °C in the presence of PAA. (**c**) $NaYF_4$ NPs synthesized through a solvothermal method in ethanol with linoleic acid and sodium linoleate as additives. Reprinted with permission from © Nanophotonics, 6, 881–921, (2017)

9.6 Miscellaneous Functionalized Nanostructures

Functionalization of the liquid media in which self-assembled monolayers are deposited to nanostructured high surface area materials has resulted in several improvements in the process and the product materials. The processing time has been reduced by three orders of magnitude. The monolayers produced from Supercritical fluids (SCFs) demonstrate the ability to "heal" over time, resulting in nearly defect-free coatings. This lack of defects allows greater coverage, and has redefined what is meant as fully dense monolayer. Depositions of covalently bound functional groups to extremely small pore spaces (i.e., <10 Å zeolites) are possible in supercritical media. Functionalizing the internal surfaces of aerogel materials is likewise enabled using super critical fluids (SCFs), and circumvents the structural collapse of such materials brought about by surface tension. The coated aerogel materials so produced are strengthened by the coating, and have been shown to withstand immersion in aqueous solutions [12].

Functionalized Peptide Nanostructures containing bioactive signals that combine bioactivity for multiple targets with biocompatibility, improve the possibility to deliver proteins, nucleic acids, drugs and cells. Their chemical design versatility leads to a variety of possible secondary, tertiary and quaternary structures through folding and hydrogen bonding. In particular, β-sheet forming peptides have demonstrated the extraordinary ability to use intermolecular hydrogen bonding for the assembly of one-dimensional nanostructures. Furthermore, the design of self-assembly peptides for targeted functions can also include their modification with other biomolecular units such as sugars, lipid components and nucleic acid monomers. These oligopeptides, whilst designed for supramolecular self-assembly, could also serve to functionally mimic large proteins. For these reasons, self-assembling, nonimmunogenic peptides project as promising new therapeutics for human disease [13].

References

1. Thangavel P, Viswanath B, Kim S (2017) Recent developments in the nanostructured materials functionalized with ruthenium complexes for targeted drug delivery to tumors. Int J Nanomed 12:2749–2758
2. Xuanyong L, Chu PK, Ding C (2010) Surface nano-functionalization of biomaterials. Mater Sci Eng R 70:275–302
3. Issa S, Azevedo C, Pires R et al (2015) Functionalization of nanostructured TiO_2 surfaces with electrodeposited strontium doped calcium phosphate and evaluation of BSA adsorption for dental implant design. J Nanomed Nanotechnol 6:344–355
4. Doty RC, Tshikhudo TR, Brust M et al (2005) Extremely stable water-soluble Ag nanoparticles. Chem Mater 17:4630–4635
5. Philipse AP, Nechifor AM, Pathmamanoharan C (1994) Isotropic and birefringent dispersions of surface modified silica rods with a boehmite-needle core. Langmuir 10:4451–4458
6. Gooding JJ, Mearns F, Yang W et al (2003) Self-assembled monolayers into the 21st century: recent advances and applications. Electroanalysis 15:81–96
7. Dharap SS, Qiu B, Williams GC et al (2003) Molecular targeting of drug delivery systems to ovarian cancer by BH3 and LHRH peptides. J Control Release 91:61–73
8. Park S, Lim JH, Chung SW et al (2004) Self-assembly of mesoscopic metal-polymer amphiphiles. Science 303:348–351
9. Mammeri F, Le Bourhis E, Rozes L et al (2005) Mechanical properties of hybrid organic-inorganic materials. Mater Chem 15:3787–3811
10. Santos C, Alves MM, Montemor MF et al (2017) Bioresorbable metallic implants: surface functionalization with nanoparticles and nanostructures. In: Advanced materials and their applications—micro to nano scale, pp 219–242
11. Escudero A, Becerro AI, Carrillo-Carrion C et al (2017) Rare earth based nanostructured materials synthesis functionalization properties and bioimaging and biosensing applications. Nanophotonics 6:881–921
12. Zemanian TS, Fryxell GE, Liu J et al (2001) Chemical functionalization of nanostructured materials using supercritical reaction media. In: Proceedings of the 1st IEEE conference on nanotechnology, pp 288–292
13. Peran M, Garcia MA, Lopez-Ruiz E et al (2012) Functionalized nanostructures with application in regenerative medicine. Int J Mol Sci 13:3847–3886

Chapter 10
Characterization and Technical Analysis of Nanostructured Materials

Abstract The morphology, structural and other properties of nanostructured materials are analyzed by different characterization techniques. In the below characterization studies, the shapes, sizes, and structures of nanostructured materials and their distribution are investigated. This chapter discuss briefly on analysis techniques.

In order to realize the correlation between structure and properties, nanocrystalline materials need to be characterized on both atomic and nanometer scales. The characteristic of above involves determining the shapes and sizes of nanoparticles and understanding of the inter-particle interactions [1]. A number of experimental techniques have been employed to yield structural information on nanocrystalline materials.

1. Atomic force microscopy (AFM)
2. Transmission electron microscopy (TEM)
3. Scanning electron microscopy (SEM)
4. Field ion microscopy (FIM)
5. Absorption spectra
6. Diffraction of X-rays, electrons or neutrons
7. Rutherford back scattering (RBS)
8. Raman spectroscopy
9. Auger electron spectroscopy (AES)
10. Photoluminescence and Photoluminescence excitation.

Techniques such as scanning electron microscopy (SEM) or transmission electron microscopy (TEM) are prominent examples of these approaches; with the latter atomic resolution is possible but sample preparation is time consuming and may introduce artefacts. Two other examples of characterization techniques based on high energy ions and electrons, which yield not only images but other, e.g. chemical information, are Nano-SIMS (secondary ion mass spectrometry) with a spatial resolution of about 50 nm, and XPS (X-ray photoelectron spectroscopy) and PEEM (photoemission electron microscopy) using synchrotron radiation, which in contrast to hard-to-focus conventional X-rays can be focused to yield spatial resolutions of 100 nm or even lower [2].

T. D. Thangadurai et al., *Nanostructured Materials*, Engineering Materials,
https://doi.org/10.1007/978-3-030-26145-0_10

10.1 Atomic Force Microscopy (AFM)

Atomic-force microscope (AFM) allows obtaining the three-dimensional images of a surface of solid state with nanoscale resolution. However in a usual mode with the help of AFM it is possible to obtain data only about surface of solid samples, because AFM cantilever tip cannot glance in deeper near-surface layers of substance. The level-by-level removal of superficial layers, for example with the help of their dissolution or chemical etching in liquid environment, allows taking off this restriction. This method is especially effective for study of nanostructured materials which have different rate of dissolution of separate nanofragments. The selective chemical etching of the surface of such sample can reveal thin structure of such substance in this case. It is essential, that as it is possible to carry out researches with AFM in liquid environment so there is an opportunity to observe transformation of the surface during etching in in situ mode in a real time scale [3].

10.2 X-Ray Diffraction (XRD)

XRD pattern from nanostructured materials is similar to that from nanocrystalline materials. If the lattice orientations of particular entities forming a solid, nanostructured material are distributed at random, the interference functions are identical. The nanostructured materials can be multiphase. The phases can all be nanocrystalline or a nanocrystalline phase can be embedded inside a polycrystalline and/or amorphous matrix. Thus besides the influence of nanocrystal size, also the grain boundaries, defects and microstrains are to be considered. Simple modeling of XRD pattern from material with different density and arrangement of dislocations in the grain boundaries show that the different methods of its analysis (Scherrer's equation, Williamson-Hall method, Warren-Averbach), based on different assumptions can bring about significantly different results. In many instances the method of material preparation and its history can suggest which factors can mostly influence the observed diffraction pattern. The measurements of XRD profiles on nanocrystalline palladium, depending on the history of the preparation of the sample shows that there is a different structure of grain boundaries and the number of atoms at the grain boundaries being not on crystal lattice side, i.e. in disordered state. A particular example of nanostructured materials can be supported metal catalysts in which the prevailing part of a mass is a support, and interest is mostly in the structure of a small fraction of the highly dispersed nanocrystalline metal. Due to the high specific surface of the support and usually low loading of the metal, the nanocrystals of the metal phase are distanced each from other. The XRD pattern of the catalyst is usually a superposition of two: from the metal phase and, usually predominant, from the support. The contribution to the pattern originating from the distances between the support and the metal in most cases can be neglected. The simplest case can be when the background is from amorphous (or quasi-amorphous) phase which can be e.g. for silica. In such a case the background can easily be modeled with smooth analytical function and subtracted from the measured pattern and the resulting profile fitted to an analytical function [4].

10.3 Scanning Probe Microscopies (SPM)

The new class of characterization techniques has to be emphasized, which allows the characterization of surfaces on the nanometer scale with respect to their topography, but also to a wide variety of other properties. These are the scanning probe microscopies (SPM), which have been rapidly developed after the invention of the scanning tunneling microscopy (STM) by Binnig and Rohrer in 1982, and subsequently that of the atomic force microscope (AFM) by Binnig et al. in 1986. All SPM techniques rely on the same principle: A probe or tip with an apex of nanometer dimensions is rastered by means of piezoelectric scanners with nanometer precision over a surface, while the interaction between tip and surface is measured. The difference between all the various techniques can be found in the type of interaction between tip and surface exploited which can rely e.g. on electrical, mechanical, optical, magnetic or thermal effects. In the original SPM tunneling currents between a conductive surface and the atoms at the very end of a sharp metal tip are measured; as a consequence, the lateral resolution is extremely high; it is possible to observe the electron distribution around single surface atoms [5].

10.4 Field Ion Microscopy (FIM)

The utility of the FIM for preparing and characterizing tips destined for SPM experiments, the main advantage of an atomically defined tip apex is that if the exact atomic arrangement of the apex is known, the electronic structure of the tip and the lateral resolution of the STM are predetermined. The same detailed knowledge of the apex termination, as well as the tip radius, is useful for atomic force microscopy (AFM) experiments and important in the interpretation of results from combined STM and AFM experiments. Largely unknown, the atomic-scale tip structure is directly responsible for image contrast, as well as the details of the measured electronic properties by scanning tunneling spectroscopy, chemical bonding forces by force spectroscopy, and yield point of materials by indentation. In order to obtain quantitative and reproducible data, which could be considered as a benchmark for computational simulations, one requires a tip with known atomic structure. For example, FIM characterized tips could act as atomically defined electrodes to build junctions to single molecules, where transport properties are sensitive to the atomic arrangement of the contact electrodes.

The FIM is the study of diffusion of single atoms or clusters on crystal surfaces, and this technique has generated most of the existing experimental data about the diffusion of atoms and clusters on metal surfaces. Another unique feature of the FIM is the possibility of integrating a time-of-flight (ToF) mass spectrometer to enable chemical analysis with single-atom sensitivity and a spatial resolution of several angstroms. The combined technique is known as atom probe field ion microscopy (AP-FIM) or more simply as the atom probe. In AP-FIM, a pulsed field is applied to the FIM specimen to field evaporate a small amount of material from its surface to be chemically analyzed in the ToF unit; it is a destructive technique, consuming the sample as data is collected. Applications of AP-FIM include the investigation of chemical segregation of elements at crystalline defects (steels and semiconductors), local studies of grain boundaries in nanocrystalline materials [with grains too small for techniques like electron backscatter diffraction (EBSD)], and short-range order in materials (e.g., in high-temperature superconductors) [6].

10.5 Raman Spectroscopy

Raman spectroscopy is a powerful tool for the characterization of nano-sized materials and structures. It is widely used for the study of phonon confinement effects, the effect of the increase in local temperature, strain and substitutional effects, lattice distortion, the presence of structural defects and nonstoichiometric in different kinds of nanomaterial. Several factors like phonon confinement, strain, non-homogeneity of the size distribution, defects and nonstoichiometric, as well as an harmonic effects due to temperature increases can contribute to the changes in the peak position, line, width and shape of the Raman modes in nanostructures [7].

The grain size and its distribution, the existence of mixed phases, the value and type of the strain, deviations from stoichiometry as well as type of stoichiometric defects, etc. also have great influence on the Raman spectra of nanomaterials. The Raman spectroscopy method can be used for the characterization of nano-powdered oxides like TiO_2, CeO_2 and ZnO, as well as nanostructured $ZnSe/SiO_x$ multilayers. The shift, broadening and asymmetric shape of the Raman modes, observed in the nanomaterials, are compared to spectra obtained from the phenomenological model, which takes into account disorder effects through the breakdown of the $k = 0$ Raman-scattering selection rule, as well as the an harmonicity, which is incorporated through the 3- and 4-phonon decay processes. The application of a three-dimensional (3D) confinement model appropriate to isolated or loosely connected nanoparticles shows that the shift and broadening of the Raman peak in some nanopowders are dominated by the strong confinement and inhomogeneous strain (CeO_2), while in the others an harmonic effects (TiO_2) or tensile strain (ZnO) plays the main role. On the other hand, the shift and asymmetric broadening of the Raman mode in nanostructured $ZnSe/SiO_x$ multilayers is analyzed by a one-dimensional (1D) confinement model, appropriate for very thin films or quantum wells. Due to the low dimensionality of these structures, the surface phonon modes are also observed. The infrared (IR) spectra of nanocrystalline solids differ from the spectra of monocrystals, due to the polycrystalline character and island structure of nanoparticles. From IR spectra, it is possible to get information about the energy gap, grain size, porosity, nature of the surface bonds, and chemical reactions occurring at the nanoparticle surface [8].

10.6 Absorption Spectroscopy (UV-Vis)

The optical absorption spectrum of ZnO nanorods and nanotubes are shown at room temperature. The UV-visible spectrum shows a strong exciton absorption peak at 366 and 362 nm for ZnO nanorods and nanotubes, respectively and blue shifted relative to the bulk exciton absorption (380 nm) as a result of the quantum size effect. According to the quantum confinement theory, the electrons in the conduction band and holes in the valance band are confined spatially by the potential barrier of the surface. Because of the confinement of electrons and holes, the optical transition energy from the valence top to the conduction bottom increases and the absorption maximum shifts to the shorter wavelength region. The stronger exciton effect is an important character of quantum confinement in nano semiconductors, in which the electrons, holes and excitons have limited space to move and their motion is possible for definite values of energies. The continuum of states in conduction and valence bands are broken down into discrete states with an energy spacing relative to band edges which is approximately inversely proportional to the square of the particle size and reduced mass. The highest occupied valence band and the lowest unoccupied conduction band are shifted to more negative and positive values respectively resulting in the widening of band gap, which leads to the effective band gap larger than its bulk value. Thus, there will be a blue shift in the absorption spectra with reduction in crystallite size. The highly confined carriers in the nanotubes enhance the coupling interaction with each other, thus the excitons bounded becomes very stronger. This is the reason why the absorption peak of ZnO nanotubes has a slight blue shift compared with that of nanorods [9].

10.7 Photoluminescence Spectroscopy (PL)

Photoluminescence (PL) spectroscopy is a useful technique for the study and characterization of materials and dynamical processes occurring in materials, specifically the optical properties of the materials. This technique involves measuring the energy distribution of emitted photons after optical excitation. This energy distribution is then analyzed in order to determine the properties of material, including defect species, defect concentrations, possible stimulated emission etc. It is widely recognized as a useful tool for characterizing the quality of semiconductor materials as well as for elucidating the physics which may accompany radiative recombination. PL is useful in quantifying (1) optical emission efficiencies, (2) composition of the material, (3) impurity content, and (4) layer thickness etc. It is well known that the radiative transition between the band-gap in semiconductors is the most important one. So the PL measurement not only helps us determine the band gap of a semiconductor especially for a new compound semiconductor, but also guides us to accomplish the band gap engineering, which is particularly significant for prompting the practical applications of a semiconductor in the industry [10].

10.8 Field Emission Scanning Electron Microscopy (FESEM)

A field emission cathode in the electron gun of a SEM provides narrower probing beams resulting in both improved spatial resolution and less sample charging. Such systems are designated as field emission scanning electron microscopes (FESEM). To achieve this increased electron focusing a different gun design is required. In this design, electrons are expelled by applying a high electric field very close to the filament tip. The size and proximity of the electric field to the electron reservoir in the filament controls the degree to which electrons tunnel out of the reservoir. One type of field emission gun commonly used is known as the Schottky in-lens thermal FESEM electron gun. Cold gun alternatives are available for even finer FESEM resolution; however, these suffer rapid degradation and can therefore lead to expensive operation due to relatively frequent placement. The field emission guns have higher stability, can allow higher current and hence provide a smaller spot size. Under good operating conditions, a typical FESEM resolution of 1 nm is achievable. Elements that add to improved operation and FESEM resolution include designs with a beam booster to maintain high beam energy, an electromagnetic multi hole beam aperture changer, a magnetic field lens and a beam path which has been designed to prevent electron beam crossover [11]. Nanostructures have been characterized by FESEM in different morphological formations including nanoflowers, nanosheets, nanoparticles and thin films. Molybdenum disulphide (MoS_2) nanosheets exhibit interesting conductive/semiconductive, magnetic, photoluminescence, and photocatalytic and field effect transistor properties. The properties of MoS_2 nanosheets depend upon the method used to generate them, and depending upon the structural properties, these nanosheets can be used for applications in optoelectronics, energy harvesting, spinelectronics, etc. [12].

10.9 Confocal Microscopy

In confocal microscopy two focusing arrangements are used to focus on the point in a sample to be imaged. One focuses laser light through an objective lens to the point of interest, and the other focuses the reflected light to the imaging sensor. Light from the point to be imaged is passed through a pinhole such that all extraneous out of focus light is removed. This allows lateral resolutions approximately 1.4 times greater than in conventional microscopes to be achieved with confocal microscopy. The depth of the focal plane depends on the specimen optical properties and importantly on the squared value of the objective lens numerical aperture. Three dimensional reconstructions of cells and surfaces can be achieved with this technique. To achieve this, the sample is scanned such that one 2D slice is recorded, the focal plane of the sample is then moved a prescribed amount where the next 2D slice is recorded and this sequence is repeated until the required volume is scanned. Image processing software is then used to process the collected data to reconstruct the 3D object. Confocal microscopes are most often used to image biological systems and semiconductor surfaces. Three variations of scanning are available in confocal microscope systems [13].

In the conventional confocal laser scanning microscope, the sample is raster scanned which results in a scanning rate of about three frames per second. Such systems provide the highest spatial resolution; however for higher temporal resolution, the spinning disc (Nipkow) and the programmable array microscope (PAM) systems can provide rates of 30 frames per second. In a Nipkow disc system, a thin disc with hundreds of spirally patterned pinholes is spun in the light path to the objective lens. The pinholes only allow perpendicularly oriented rays of light to penetrate which allows high scanning speeds independent of the laser scanning speed. PAM is a variation on this whereby an acousto- or electro-optical filter can be patterned to automatically produce the pinhole pattern required. Such a system can allow for up to 1000 beams to simultaneously scan the entire field at millisecond scan speeds. High-frequency scanning has the added advantage of reducing exposure of sensitive samples to photons which may cause damage due to photobleaching or phototoxicity [14].

10.10 Transmission Electron Microscope (TEM)

TEM is an established characterization technique, which can provide both image mode and diffraction mode information from a single sample. It is regarded as one of the main techniques for nanomaterial characterization, largely due to its high lateral spatial resolution in the region of 0.08 nm. A feature of nanomaterials is that specific properties, for example, colour, can be related to a particle size. Agglomeration of nanoparticles or failure to isolate individual nanostructures is likely to result in anomalous property characterization. Characterizing the elastic or mechanical properties of individual nanoparticle is a challenge to many existing testing and measurement techniques. It is difficult to pick up samples and difficult to clamp samples, in order to test for tensile strength or creep, for example [15].

The modern TEM are capable of formatting nanometer size electron probes having diameters ranging from 2 to 5 nm. This formation is possible by employing a multistage condenser lens system. This lens system makes scanning transmission mode possible, and the resulting electron probe diameter defines the resolution of the system. Therefore, in addition to thin samples, specimens with higher degree of crystallinity and thickness can be imaged by TEM. Multistage condenser lens systems enable recording of secondary and backscattered electrons. This has advantages for imaging thick or crystalline specimens and for recording secondary electrons and backscattered electrons. The inhomogeneity in cathode luminescence can also be recorded using complex multistage condenser lens system for correlation with structural defects. Cathodoluminescence microscopy is a useful characterization technique in various fields related to optoelectronics, energy, geology, cellular biology and healthcare. Traditionally scanning electron microscopy (SEM) has been used to study the cathodoluminescence of bulk samples as well as nanoparticles. The limited resolution of SEM up to 20 nm in the most advanced systems restricts the use of SEM for microstructure correlation with the cathodoluminescence. Therefore high-resolution cathodoluminescence microscopy is possible using TEM [16].

10.11 X-Ray Photoelectron Spectroscopy (XPS)

When excess electromagnetic energy is transferred to an electron that is in an outer shell, it is called an Auger electron. An analysis of these electrons for chemical identification is known as Auger electron spectroscopy (AES). X-ray photoelectron spectroscopy (XPS) analyses electron emission of similarly high energy. XPS can be used to measure the chemical or electronic state of surface elements, detect chemical contamination or map chemical uniformity of biomedical implant surfaces. For XPS the material to be examined is irradiated with aluminium or magnesium X-rays. Monochromatic aluminium Kα X-rays are normally produced by diffracting and focusing a beam of non-monochromatic X-rays off of a thin disc of crystalline quartz. Such X-rays have a wavelength of 8.3386 Å, corresponding photon energy of 1486.7 eV, and provide a typical energy resolution of 0.25 eV. Non-monochromatic magnesium X-rays have a wavelength of 9.89 Å, corresponding photon energy of 1253 eV and a typical energy resolution of 0.90 eV. The kinetic energy of the emitted electrons is recorded. This kinetic energy of the ejected electrons is directly related to the element-specific atomic binding energy of the liberated. A plot of these energies against the corresponding number of electron counts provides the spectrum which indicates the qualitative and quantitative elemental composition [17]. At these higher energies, XPS only analyses to a depth of 10 nm into the surface. Electrons emitted at greater depths are recaptured or trapped in various excited states within the material. Spectral profiles up to 1 μm deep can however be obtained by continuous spectral recording during ion etching or from consecutive ion etching and XPS measurement steps.

XPS is usually performed in UHV and typically provides resolutions down to 1000 ppm. With optimum settings and long recording times, resolutions down to 100 ppm can be achieved. Non-monochromatic X-ray sources can produce a significant amount of heat (up to 200 °C) on the surface of the sample as the anode producing the X-rays is typically only a few centimeters from the sample. This level of heat when combined with high-energy Bremsstrahlung X-rays can degrade the surface. Organic chemicals are therefore not routinely analyzed by non-monochromatic X-ray sources [18].

10.12 Auger Electron Spectroscopy (AES)

When an electron or ion is incident on a semiconductor, it may transfer enough energy to an inner shell electron to eject it from its parent atom. The atom is in an excited state, and, to lower its energy, an electron from a less tightly bound shell may fill the hole while simultaneously emitting a third electron from the atom. This ejected atom is known as an Auger electron. Its energy is related specifically to the electron energy levels involved in the process and, therefore, is characteristic of the atom concerned. Since the Auger process is not a three-electron process, hydrogen nor helium can be detected since both have less than three electrons [19].

AES has two distinct advantages over EDX analysis. It is a far more surface-sensitive technique. Escape depths range from less than a nanometer to a few nanometers. In EDX it can be difficult to analyze small particles on a substrate, because the electron beam passes through the particles and spreads out in the substrate below it. There is the potential for chemical-state information in Auger spectroscopy, for example, the oxidation state of silicon at a Si–SiO_2 interface may be ascertained. EDX does not provide chemical state information. AES has found applications in measuring semiconductor composition, oxide film composition, silicide's, metallization, particle analysis and the effects of surface cleaning. AES measurements are made in a high vacuum environment (10^{-12}–10^{-10} torr) to retard the formation of hydrocarbon contamination layers on the sample surface. Scanning Auger microscopy (SAM) allows surfaces to be mapped for one selected element at a time. In this mode the electron beam is scanned over a selected area. The Auger intensity is measured at each point of the area. SAM requires higher beam currents and is much slower than SEM/EDX [20].

References

1. Ramrakhiani M (2012) Nanostructures and their applications. Recent Res Sci Technol 4:14–19
2. Winhold M, Leitner M, Lieb A et al (2017) Correlative in-situ AFM & SEM & EDX analysis of nanostructured materials. Microsc Microanal 23:26–27
3. Bukharaev AA, Nurgazizov N, Mozhanova AA et al (1999) Atomic force microscopy characterization of nanostructured materials using selective chemical etching. Nanostruct Phys Technology 7:236–239
4. Knauth P, Schoonman J (2004) Nanostructured materials selected synthesis methods, properties and applications. Kluwer Academic Publishers, New York
5. Fink HW (1986) Mono-atomic tips for scanning tunneling microscopy. IBM J Res Div 30:460–465

6. Paul W, Grutter P (2015) Field ion microscopy for the characterization of scanning probes. In: Surface science tools for nanomaterials characterization. Springer, Berlin, pp 159–198
7. Scepanovic MJ, Grujic-Brojcin M, Dohcevic-Mitrovic Z et al (2007) Vibrational spectroscopy methods in the characterization of nanostructured materials. JOAM 9:30–36
8. Zhu KR, Zhang MS, Chen Q et al (2005) Size and phonon-confinement effects on low frequency Raman mode of anatase TiO_2 nanocrystal. Phys Lett A 340:220–227
9. Beshkar F, Amiri O, Salehi Z (2017) Synthesis of $ZnSnO_3$ nanostructures by using novel gelling agents and their application in degradation of textile dye. Sep Purif Technol 184:66–71
10. Schmitt SW, Sarau G, Christiansen S (2015) Observation of strongly enhanced photoluminescence from inverted cone-shaped silicon nanostructures. Sci Rep 5:17089
11. Jusman Y, Ng SC, Abu Osman NA (2014) Investigation of CPD and HMDS sample preparation techniques for cervical cells in developing computer-aided screening system based on FE-SEM/EDX. Sci World J 289817:1–11
12. Sall T, Mollar M, Mari B (2016) Substrate influences on the properties of SnS thin films deposited by chemical spray pyrolysis technique for photovoltaic applications. J Mater Sci 51:7607–7613
13. Chu KKW, Chen JS, Der Chang L et al (2017) Graphene-edge probes for scanning tunneling microscopy. Optik 130:976–980
14. Fulwyler M, Hanley QS, Schnetter C et al (2005) Selective photoreactions in a programmable array microscope (PAM): Photoinitiated polymerization, photodecaging, and photochromic conversion. Cytometry A 67:68–75
15. Brundl CR, Evans CA, Wilson S (1992) Encyclopedia of materials characterization: surfaces, interfaces, thin films. Gulf Professional Publishing, Texas
16. Barbin V (2013) Application of cathodoluminescence microscopy to recent and past biological materials: a decade of progress. Mineral Petrol 107:353–362
17. Matthew J (2004) Surface analysis by Auger and x-ray photoelectron spectroscopy. In: Briggs D, Grant JT (eds). IMPublications, Chichester, UK and Surface Spectra, Manchester, UK, 2003, 900 pp. ISBN 1-901019-04-7
18. Ahad IU, Budner B, Fiedorowicz H et al (2013) Nitrogen doping in biomaterials by extreme ultraviolet (EUV) surface modification for biocompatibility control. Eur Cell Mater 26:145–146
19. Jenkins TE (1998) Semiconductor science; growth and characterization techniques. Prentice Hall, Harlow, Essex
20. Gao XL, Pan JS, Hsu CY (2006) Laser-fluoride effect on root demineralization. J Dent Res 85:919–923

Chapter 11
Fabrication of Nanostructures

Abstract This chapter explains the techniques that are available for fabricating nanostructures and also explain the substrates and wafers, modification of materials, lithography, film deposition, wet and dry etching, wafer bonding and packaging.

With a bottom-up approach, nanostructures are formed molecule by molecule, using methods such as chemical vapor deposition or self-assembly. By contrast, top-down fabrication can be likened to sculpting from a base material, and typically involves steps such as deposition of thin films, patterning, and etching [1].

11.1 Lithography

Lithography is referred as photoengraving, and is the process of transferring a pattern into a reactive polymer film, termed as resist, which will subsequently be used to replicate that pattern into an underlying thin film or Many techniques of lithography have been developed in the last half a century with various lens systems and exposure radiation sources including photons, X-rays, electrons, ions and neutral atoms. In spite of different exposure radiation sources used in various lithographic methods and instrumental details, they all share the same general technical approaches and are based on similar fundamentals [2].

11.1.1 Photolithography

Typical photolithographic process consists of producing a mask carrying the requisite pattern information and subsequently transferring that pattern, using some optical technique into a photoactive polymer or photoresist. There are two basic photolithographic approaches: (i) shadow printing, and proximity printing, (ii) projection printing. Figure 11.1 outlines the basic steps of the photolithographic process,

T. D. Thangadurai et al., *Nanostructured Materials*, Engineering Materials,
https://doi.org/10.1007/978-3-030-26145-0_11

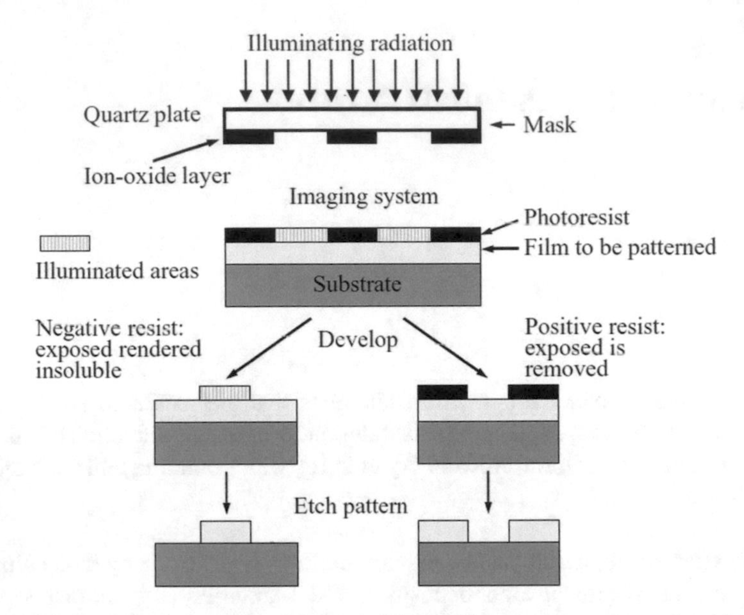

Fig. 11.1 Schematic representation of the photolithographic process sequences, in which images in the mask are transferred to the underlying substrate surface. Reprinted with permission from © Journal of Nanomaterials, Article. No. 486301, p. 21 (2012)

in which the resist material is applied as a thin coating over some base and subsequently exposed in an image-wise fashion through a mask, such that light strikes selected areas of the resist material. The exposed resist is then subjected to a development step. Depending on the chemical nature of the resist material, the exposed areas may be rendered more soluble in some developing solvent than the unexposed areas, thereby producing a positive tone image of the mask. Conversely, the exposed areas may be rendered less soluble, producing a negative tone image of the mask. The effect of this process is to produce a three-dimensional relief image in the resist material that is a replication of the opaque and transparent areas of the mask. The areas of resist that remain following the imaging and developing processes are used to mask the underlying substrate for subsequent etching or other image transfer steps. The resist material resists the etchant and prevents it from attacking the underlying substrate in those areas where it remains in place after development. Following the etching process, the resist is removed by stripping to produce a positive or negative tone relief image in the underlying substrate. Diffraction sets the limit of the maximum resolution or the minimum size of the individual elements by photolithography, which can be obtained [3].

Diffraction refers to the apparent deviation of light from rectilinear propagation as it passes an obstacle such as an opaque edge and the phenomenon of diffraction can be understood qualitatively as follows. According to geometrical optics, if an opaque object is placed between a point light source and a screen, the edge of the object will cast a sharp shadow on the screen. No light will reach the screen at points within the geometrical shadow, whereas outside the shadow the screen will be uniformly illuminated. In reality, the shadow cast by the edge is diffuse, consisting of alternate bright and dark bands that extend into the geometrical shadow. This apparent bending of light around the edge is referred to as diffraction, and the resulting intensity distribution is called a diffraction pattern. Obviously diffraction causes the image of a perfectly delineated edge to become blurred or diffused at the resist surface. The theoretical resolution capability of shadow photolithography with a mask consisting of equal lines and spaces of width b is given by:

$$2b_{\min} = 3\sqrt{\lambda(S + d/2)}$$

where 2b is the grating period (1/2b is the fundamental spatial frequency ν), s the gap width maintained between the mask and the photoresist surface, λ the wavelength of the exposing radiation and d the photoresist thickness. For hard contact printing, s is equal to 0, and from the equation, the maximum resolution for 400 nm wavelength light and a 1 pm thick resist film will be slightly less than 1 km [4].

11.1.2 Contact-Mode Photolithography (CMP)

In contact-mode photolithography, the mask and wafer are in intimate contact, and thus this method can transfer a mask pattern into a photoresist with almost 100% accuracy and provides the highest resolution. Other photolithographic techniques can approach but not exceed its resolution capabilities. However, the maximum resolution is seldom achieved because of dust on substrates and non-uniformity of the thickness of the photoresist and the substrate. Such problems can be avoided in proximity printing, in which, a gap between the mask and the wafer is introduced. However, increasing the gap degrades the resolution by expanding the penumbral region caused by diffraction. The difficulties in proximity printing include the control of a small and very constant space between the mask and wafer, which can be achieved only with extremely flat wafers and masks [5]. Projection printing differs from shadow printing. In projection printing techniques, lens elements are used to focus the mask image onto a wafer substrate, which is separated from the mask by many centimeters. Because of lens imperfections and diffraction considerations, projection techniques generally have lower resolution capability than that provided by shadow printing. The resolution limit in conventional projection photolithography is determined largely by the well-known Rayleigh's equation. The resolution, i.e. the minimum resolvable feature, R, and the corresponding depth of focus (DOF) are given by the following:

$$R = \frac{k_1 \lambda}{NA}$$

$$DOF = \frac{k_2 \lambda}{NA^2}$$

Here A is the exposure wavelength, k_1 and k_2 are constants that depend on the specific resist material, process technology and image-formation techniques used, and NA is the numerical aperture of the optical system and is defined as

$$NA = n \, Sin \, \theta$$

where n is the index of refraction in image space and is usually equal to 1 (air or vacuum), and θ is the maximum cone angle of the exposure light beam. The diffraction limit is a very basic law of physics directly related to Heisenberg's uncertainty relation. It restricts any conventional imaging process to a resolution of approximately λ/2 [6].

11.1.3 Deep Ultra-Violet Lithography (DUV)

The Deep Ultra-Violet Lithography is based on exposure at wavelengths below 300 nm, presents far more difficult technical challenges. Classical UV sources have lower output power in the DUV Excimer lasers can provide 10–20 W of power at any one of several wavelengths in the DUV. Of particular interest are the KrCl and KrF excimer lasers, which have outputs at 222 and 249 nm, respectively. High intensity, microwave powered emission sources provide substantially higher DUV output than classical electrode discharge mercury lamps Light sources with shorter wavelengths exploited for optical lithography include: KrF excimer laser with a wavelength of 249 nm, ArF excimer laser of 193 nm, F_2 excimer laser of 157 nm. Optical lithography allows one to obtain patterns with a minimal size of ~100 nm. Extreme UV lithography with wavelengths in the range of 11–13 nm has also been explored for fabricating features with even smaller dimensions and is a strong candidate for achieving dimensions of 70 nm and below [7].

11.1.4 Phase-Shifting Photolithography

In this method, a transparent mask induces abrupt changes of the phase of the light used for exposure, and cause optical attenuation at desired locations. These phase masks, also known as phase shifters, have produced futures of ~100 nm in photoresist. Figure 11.2 schematically illustrates the principles of phase-shifting lithography. A clear film, i.e. a phase shifter or a phase mask, whose thickness is λ/2 (n − 1) is placed on a photoresist with conformal contact, the phase angle of the exposure light passing through the film shifts by the amount of π to the incident light arriving at the surface of the photoresist. Here λ is the wavelength of the exposure light and n is the index of refraction of the phase mask. Because the light phase angle between the phase shifter and the photoresist is inverted, the electric field at the phase shifter edge is 0. So the intensity of the exposure light at the surface of the photoresist would

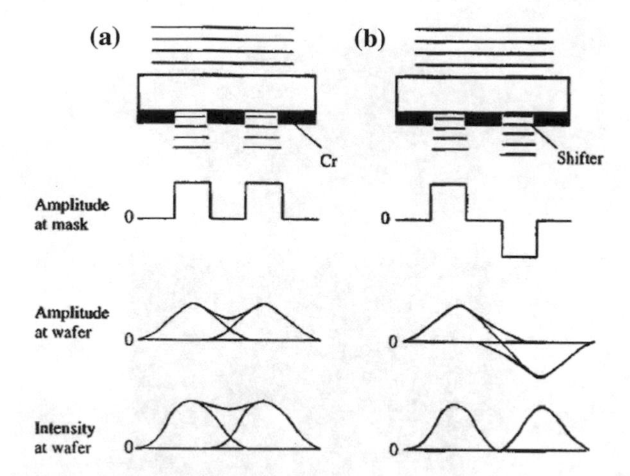

Fig. 11.2 Schematic illustrating the principles of phase shifting lithography, which utilizes the optical phase, changes at the phase shifter edge

be zero. An image having zero intensity can be formed about the edge of the phase shifter. Phase masks can be used in both projection and contact-mode photolithographic techniques. For a phase-shifting contact-mode photolithography, there are two possible approaches to increase the resolution: (i) reducing the wavelength of the source of exposure light and (ii) increasing the index of refraction of the photoresist. The achievable photolithographic resolution is roughly of ~λ/4n, where A is the wavelength of the exposure light and n is the refractive index of the photoresist. Although contact-mode photolithography with a phase-shifting mask has a higher resolution, conformal contact between the phase-shifting mask and the photoresist on wafer is difficult to achieve, due to the presence of dust, non-uniformities in the thickness of the photoresist, and bowing of the mask or the substrate. However, by introducing elastomeric phase-shifting masks, conformal contact can be relatively easily achieved and feature lines as narrow as 50 nm have been generated. The resolution achieved corresponds approximately to λ/5. An improved approach to conformal near field photolithography is to use masks constructed from "soft" organic elastomeric polymer [8]. Figure 11.3 shows a pattern created using such a contact-mode phase-shifting photolithographic process.

11.1.5 Electron Beam Lithography

A finely focused beam of electrons can be deflected accurately and precisely over a surface. When the surface is coated with a radiation sensitive polymeric material, the electron beam can be used to write patterns of very high resolution. The first experimental electron beam writing systems were designed to take advantage of the high resolution capabilities in the late sixties. Electron beams can be focused to

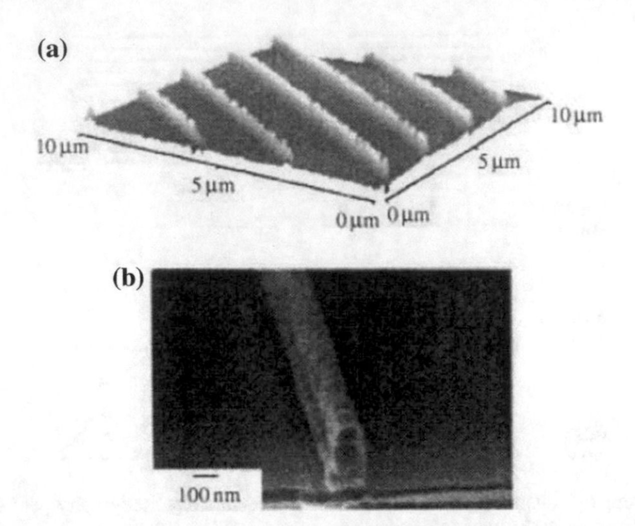

Fig. 11.3 Parallel lines formed in photoresist using near field contact-mode photolithography have widths on the order of 100 nm and are −300 nm in height as imaged by **a** AFM and **b** SEM. Reprinted with permission from © Journal of Nanomaterials, Article. No. 936876, p. 7 (2015)

a few nanometers in diameter and rapidly deflected either electromagnetically or electrostatically. Electrons possess both particle and wave properties; however, their wavelength is on the order of a few tenths of angstrom, and therefore their resolution is not limited by diffraction considerations. Resolution of electron beam lithography is, however, limited by forward scattering of the electrons in the resist layer and back scattering from the underlying substrate. Nevertheless, electron beam lithography is the most powerful tool for the fabrication of feathers as small as 3–5 nm. When an electron beam enters a polymer film or any solid material, it loses energy via elastic and inelastic collisions known collectively as electron scattering [9]. Elastic collisions result only in a change of direction of the electrons, whereas inelastic collisions lead to energy loss. These scattering processes lead to a broadening of the beam, i.e. the electrons spread out as they penetrate the solid producing a transverse or lateral electron flux normal to the incident beam direction, and cause exposure of the resist at points remote from the point of initial electron incidence, which in turn results in developed resist images wider than expected. The magnitude of electron scattering depends on the atomic number and density of both the resist and substrate as well as the velocity of the electrons or the accelerating voltage. Exposure of the resist by the forward and backscattered electrons depends on the beam energy, film thickness and substrate atomic number. As the beam energy increases, the energy loss per unit path length and scattering cross-sections decreases. Thus the lateral transport of the forward scattered electrons and the energy dissipated per electron decrease while the lateral extent of the backscattered electrons increases due to the increased electron range. As the resist film thickness increases, the cumulative effect of the small angle collisions by the forward scattered electrons increases. Thus the area exposed by the scattered electrons at the resist substrate interface is larger in thick films than in thin films. Proper exposure requires that the electron range in the polymer film be greater than the film thickness in order to ensure exposure of the resist at the interface. As the substrate atomic number increases, the electron reflection coefficient increases which in turn increases the backscattered contribution. Electron beam systems can be conveniently considered in two broad categories: those using scanned, focused electron beams which expose the wafer in serial fashion, and those projecting an entire pattern simultaneously onto a wafer. Scanning beam systems can be further divided into Gaussian or round beam systems and shaped beam systems. All scanning beam systems have four typical subsystems: (i) electron source (gun), (ii) electron column (beam forming system), (iii) mechanical stage and (iv) control computer which is used to control the various machine subsystems and transfer pattern data to the beam deflection systems [10].

11.1.6 X-Ray Lithography

X-rays with wavelengths in the range of 0.04–0.5 nm represent another alternative radiation source with potential for high-resolution pattern replication into polymeric resist materials. X-ray lithography was first demonstrated that to obtain high-resolution patterns using X-ray proximity printing by Spears and Smith. The essential ingredients in X-ray lithography include:

(1) A mask consisting of a pattern made with an X-ray absorbing material on a thin X-ray transparent membrane,
(2) An X-ray source of sufficient brightness in the wavelength range of interest to expose the resist through the mask, and
(3) An X-ray sensitive resist material.

There are two X-ray radiation sources: (i) electron impact and (ii) synchrotron sources. Conventional electron impact sources produce a broad spectrum of X-rays, centered about a characteristic line of the material, which are generated by bombardment of a suitable target material by a high energy electron beam. The synchrotron or storage ring produces a broad spectrum of radiation stemming from energy loss of electrons in motion at relativistic energies. This radiation is characterized by an intense, continuous spectral distribution from the infrared to the long wavelength X-ray region. It is highly collimated and confined near the orbital plane of the circulating electrons, thereby requiring spreading in the vertical direction of moving the mask and wafer combination with constant speed through the fan of synchrotron radiation. Synchrotrons offer the advantage of high power output. Absorption of an X-ray photon results in the formation of a photoelectron which undergoes elastic and inelastic collisions within the absorbing material producing secondary electrons which are responsible for the chemical reactions in the resist film. The range of the primary photoelectrons is on the order of 100–200 nm [11].

11.1.7 Focused Ion Beam (FIB) Lithography

The focused ion beam has been rapidly developed into a very attractive tool in lithography, etching, deposition, and doping. Since scattering of ions in the MeV range is several orders of magnitude less than that for electrons, ion beam lithography has long been recognized to offer improved resolution. The commonly used FIB_S are Ga and Au–Si–Be alloys LMI sources due to their long lifetime and high stability. FIB lithography is capable of producing electronic devices with submicrometer dimension. The advantages of FIB lithography include its high resist exposure sensitivity, which is two or more orders of magnitude higher than that of electron beam lithography, and its negligible ion scattering in the resist and low back scattering from the substrate. However, FIB lithography suffers from some drawbacks such as lower throughput and extensive substrate damage. Therefore, FIB lithography is more likely to find applications in fabricating devices where substrate damage is not critical. FIB etching includes physical sputtering etching and chemical assisted etching. Physical sputtering etching is straightforward and is to use the highly energetic ion beams to bombard the area to be etched and to erode material from the sample. The advantages of this process are simple, capable of self-alignment, and applicable to any sample material [12]. Chemical etching is based on chemical reactions between the substrate surface and gas molecules adsorbed on the substrate. Chemical etching offers several advantages: an increased etching rate, the absence of redeposition and little residual damage. Particularly, the chemical assisted etching rate ranges 10–100 folds for various combinations of materials and etchant gases, and the absence of redeposition permits very high aspect ratios. FIB can also be used for depositing. Similar to etching, there are direct deposition and chemical assisted deposition. Direct deposition uses low energy ions, whereas chemical assisted deposition relies on chemical reactions between the substrate surface and molecules adsorbed on the substrate. FIB lithography offers several advantages for the fabrication and processing of magnetic nanostructures in comparison with electron beam lithography. Ions are substantially heavier than electrons, and thus the FIB is much less influenced by magnetic properties of the material. Another advantage is its ability of achieving direct etching and/or deposition without using extra patterning steps [13].

11.1.8 Neutral Atomic Beam Lithography

In neutral atomic beams, no space charge effects make the beam divergent; therefore, high kinetic particle energies are not required. Diffraction is no severe limit for the resolution because the de Broglie wavelength of thermal atoms is less than 1 angstrom. These atomic beam techniques rely either on direct patterning using light forces on atoms that stick on the surface, or on patterning of a special resist. Interaction between neutral atoms and laser light has been explored for various applications, such as reduction of the kinetic-energy spread into the nanokelvin regime, trapping atoms in small regions of space or manipulation of atomic trajectories for focusing and imaging [14]. Basic principle of atomic beam lithography with light forces can be understood in a classical model as follows. The induced electric dipole moment of an atom in an electromagnetic wave can be resonantly enhanced by tuning the oscillation frequency of the light ω_L close to an atomic dipole transition with frequency ω_A Depending on the sign of the detuning $\delta = \omega_L - \omega_A$ the dipole moment is in phase ($\delta < 0$) or out of phase ($\delta > 0$). In an intensity gradient, this induced dipole feels a force towards the local minimum ($\delta < 0$) or maximum ($\delta > 0$) of the spatial light intensity distribution.

Therefore, a standing light wave acts as a periodic conservative potential for the motion of the atoms and forms the analogue of an array of cylindrical lenses [15]. If a substrate is positioned at the focal plane of this lens array, a periodic structure is written onto the surface. Figure 11.4 schematically illustrates the basic principles of neutral atom lithography with light forces and Fig. 11.5 shows the resulting chromium nanowires of 64 nm on silicon substrate grown by neutral atomic beam deposition with laser forces.

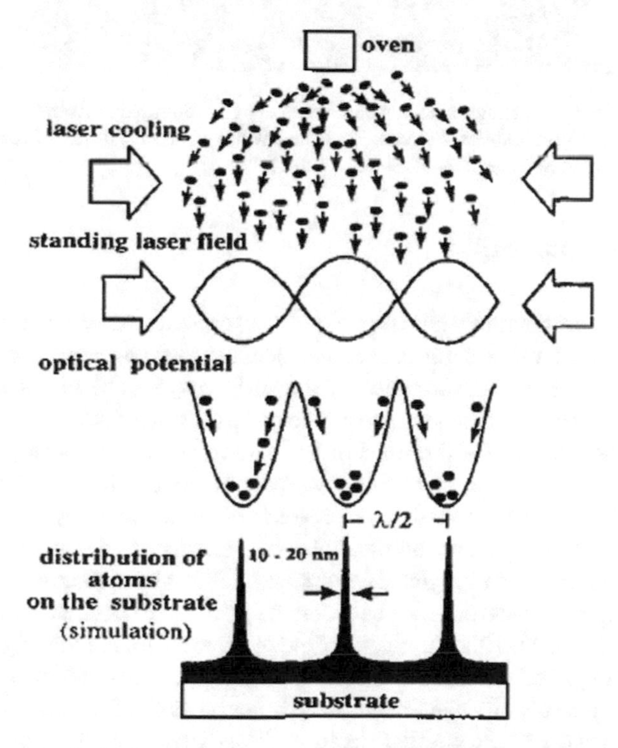

Fig. 11.4 Schematic illustrating the basic principles of neutral atom lithography with light forces. Reprinted with permission from © Proceedings of the National Academy of Sciences, 99, 6509–6513 (2002)

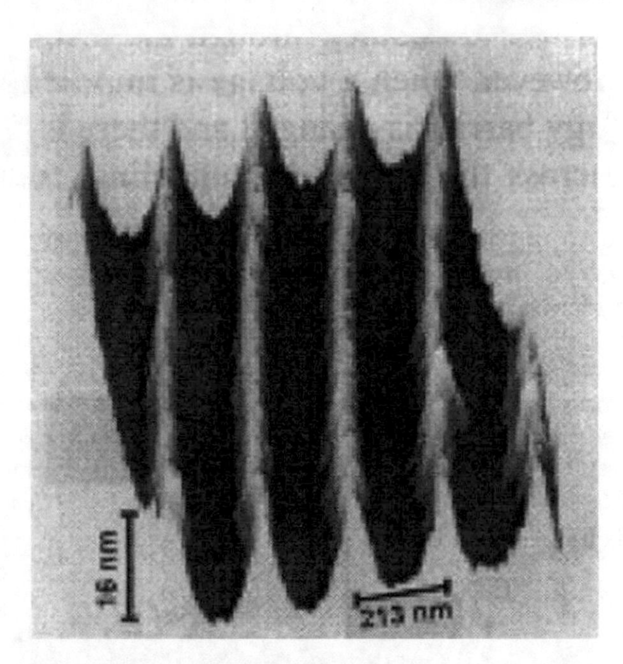

Fig. 11.5 SEM image showing chromium nanowires of 64 nm on silicon substrate grown by neutral atomic beam deposition with laser forces. Reprinted with permission from © Proceedings of the National Academy of Sciences, 99, 6509–6513 (2002)

11.2 Nanolithography

SPM-based nanolithography has been exploited for local oxidation and passivation, localized chemical vapor deposition, electrodeposition, mechanical contact of the tip with the surface, and deformation of the surface by electrical pulses. There are direct anodic oxidation of the sample surface and exposure of electron resist. Patterns with a minimal size of 10–20 nm105 or to 1 nm in UHV have been demonstrated. Nanometer holes can be formed using low energy electrons from a STM tip when a pulsed electric voltage is applied at the presence of sufficient gas molecules between the substrate and the tip. For example, holes of 7 nm deep and 6 nm wide on HOPG substrate were formed in nitrogen at a pressure of 25 bar by applying a −7 V pulse to the tip for 130 ms with the distance between the tip and the substrate being 0.6–1 nm. A possible mechanism is that the electric field induces the ionization of gas molecules near the STM tip, and accelerates the ions towards the substrate. Ions bombard the substrate and consequently nanometer-sized holes are created. A certain electric field is required to generate field emitted electrons. The diameter of electron beam ejected from a STM tip is dependent on the applied bias voltage and the diameter of the tip. At low bias (<12 V), the diameter of the ejected electron beam remains almost constant; however, the beam diameter changes significantly with bias voltage and the diameter of the tip. Nanostructures can be created using field evaporation by applying bias pulses to the STM tip-sample tunneling junction. For example, nano-dots, lines, and corrals of gold on a clean stepped Si (111) surface were fabricated by applying a series of bias pulses <10 V and ~30 ps to a STM gold tip at UHV a base pressure of $\sim 10^{-10}$ mbar. Nano-dots with diameter as small as a few nanometers can be realized. By decreasing the distance between adjacent nano-dots, it was possible to create continuous nano-lines of a few nanometer wide and over a few hundred nanometer long. A nano-corral of about 40 nm in diameter formed by many Au nano-dots for a few nanometers in diameter each was also created on the silicon (111) surface [16].

11.2.1 AFM Based Nanolithography

Direct contacting, writing, or scratching is referred to as a mechanical action of the AFM tip that is used as a sharply pointed tool in order to produce fine grooves on sample surfaces. Although direct scratching creates grooves with high precision, low quality results are often obtained due to tip wear during the process. An alternative approach is to combine scratching on a soft resist polymer layer, such as PMMA or polycarbonate, as a mask for the etching process and subsequent etching to transfer the pattern to the sample surface. This method ensures reduced tip damage, but also precludes an accurate alignment to the structures underneath. A two-layer mask has been investigated as a further improvement [17]. For example, a mask coating consisting of a thin layer of polycarbonate of 50–100 m and a film of easy-to-deform and fusible metal such as indium or tin was used to create 50 nm wide structures. Figure 11.6 is the typical layout of the sample and the process steps with AFM lithography.

11.2.2 Soft Lithography

Soft lithography is a set of non-photolithographic techniques for microfabrication that are based on the printing of SAMs and molding of liquid precursors. Soft lithography techniques include contact printing, micromolding in capillaries, microtransfer molding and replica molding. Soft lithography has been developed as an alternative to photolithography and a replication technology for both micro- and nanofabrication [18].

11.2.3 Microcontact Printing

Microcontact printing is a technique that uses an elastomeric stamp with relief on its surface to generate patterned SAMs on the surface of both planar and curved substrate. The procedure of microcontact printing is experimentally simple and inherently parallel.

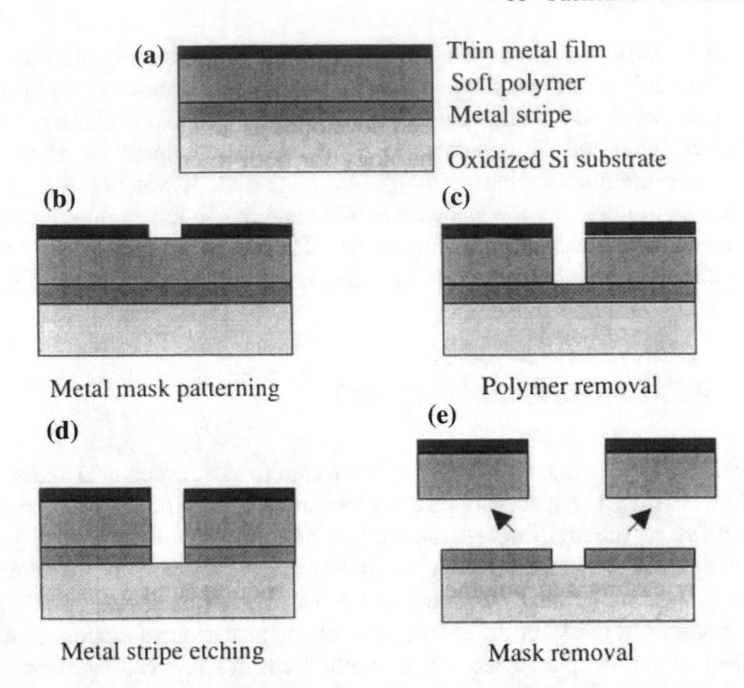

Fig. 11.6 Layout of the sample and the process steps: **a** sample multilayer structure, **b** thin mask patterning by AFM lithography, **c** polymer removal in plasma oxygen, **d** titanium stripe etching, and **e** resulting electrodes after sacrificial layers removal. Reprinted with permission from © Nanotechnology, 10, 458–463 (1999)

The elastomeric stamp is fabricated by casting and polymerizing PDMS monomer in a master mold, which can be prepared by photolithography or other relevant techniques. The stamp with a desired pattern is brought in contact with ink, a solution to form a SAM on the surface of the stamp. The inked stamp then contacts a substrate and transfers the SAM onto the substrate surface with patterned structure. A very important advantage that the microcontact printing offers over other patterning techniques is the capability to fabricate a patterned structure on a curved surface. The success of microcontact printing relies (i) on the conformal contact between the stamp and the surface of the substrate, (ii) on the rapid formation of highly ordered monolayers as a result of self-assembly, and (iii) on the autophobicity of the SAM, which effectively blocks the reactive spreading of the ink across the surface. Microcontact printing has been used with a number of systems including SAMs of alkanethiolate on gold, silver and copper, and SAMs of alkylsiloxanes on HO-terminated surfaces. Microcontact printing can routinely form patterns of alkanethiolate SAMs on gold and silver with in-pane dimensions at the scale of ~500 nm. But smaller futures, such trenches in gold as ~35 nm wide and separated by ~350 nm can be fabricated with a combination

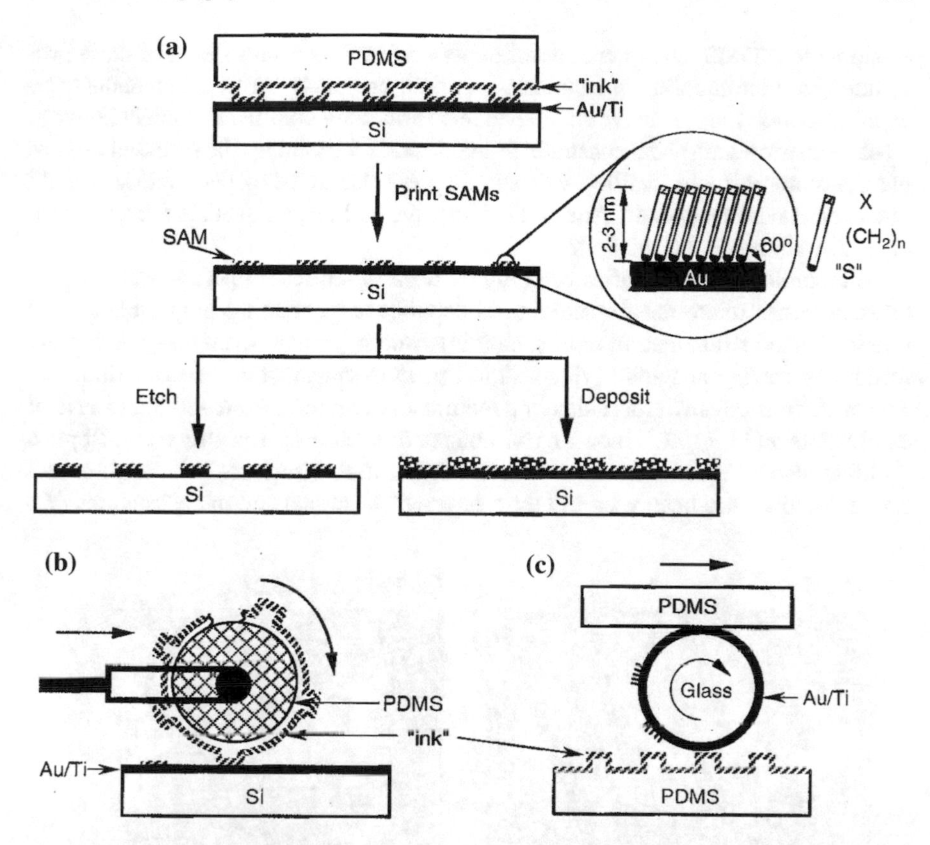

Fig. 11.7 Schematic showing the principal procedures of a typical microcontact printing: **a** printing on a planar substrate with a planar PDMS stamp, **b** printing on a planar substrate with a rolling stamp, and **c** printing on a curved substrate with a planar stamp. Reprinted with permission from © Angew Chem Int Ed Engl, 37, 550–575 (2016)

of microcontact printing of alkanethiolate SAMs and wet etching. Figure 11.7 is the schematic showing of the principal procedures of a typical microcontact printing: (a) printing on a planar substrate with a planar PDMS stamp, (b) printing on a planar substrate with a rolling stamp, and (c) printing on a curved substrate with a planar stamp [19].

11.2.4 Molding

A number of molding techniques have been developed for the fabrication of microstructures, but are also capable of fabricating nanostructures. These techniques include micromolding in capillaries, microtransfer molding, and replica molding. An elastomeric (PDMS) stamp with relief on its surface is central to each of these procedures. In micromolding in capillaries, a liquid precursor wicks spontaneously by capillary action into the network of channels formed by conformal contact between an elastomeric stamp and a substrate. In microtransfer molding, the recessed regions of an elastomeric mold are filled with a liquid precursor, and the filled mold is brought into contact with a substrate. After solidifying, the mold is removed, leaving a micro- or nanostructure on the substrate.

Micromolding in capillaries can only be used to fabricate interconnected structures, whereas microtransfer molding is capable of generating both isolated and interconnected structures. In replica molding, micro- or nanostructures are directly formed by casting and solidifying a liquid precursor against an elastomeric mold. This method is effective for replicating feature sizes ranging from ~30 nm to several centimeters and Fig. 11.8 shows AFM images of such prepared structures. Replica molding also offers a convenient route to fabricating structures with high aspect ratios. Molding has been used to fabricate microstructures and nanostructures of a

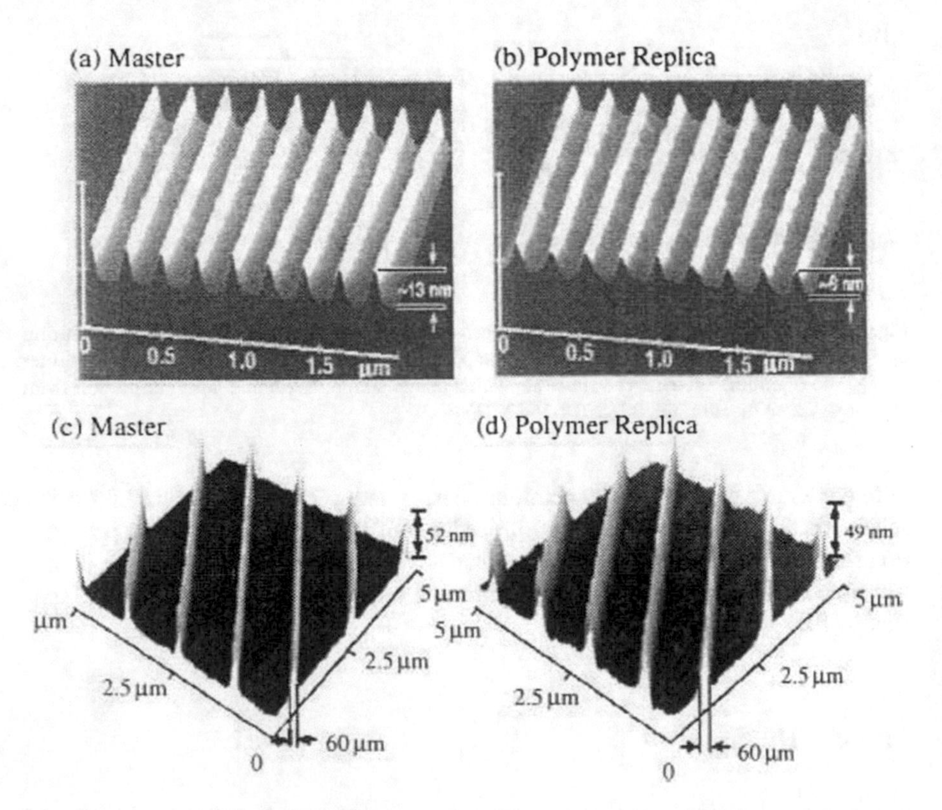

Fig. 11.8 **a** and **b** AFM images of chromium nanostructures on a master, and a polyurethane replica prepared from a PDMS mold cast from this master. **c** and **d** AFM images of Gold nanostructures on another master, and a polyurethane replica produced from a different PDMS mold cast from this master. Reprinted with permission from © Annu. Rev. Mater. Sci, 28, 153–84 (1998)

wide range of materials, including polymers, inorganic and organic salts, sol-gels, polymer beads, and precursor polymers to ceramics and carbon [20].

11.2.5 Nanoimprint

Nanoimprint lithography is a conceptually straightforward method in fabrication of patterned nanostructures. Nanoimprint lithography has demonstrated both high resolution and high throughput for making nanometer scale structures. Figure 11.9 schematically illustrates the principal steps of a typical nanoimprint process. First a stamp with the desired features is fabricated, for example by optical or electron beam lithography followed by dry etching or reactive ion etching. The material to be printed, typically a thermoplastic polymer, is spun onto a substrate where the nanostructures are to be fabricated, The second step is to press the stamp on the polymer layer with the temperature raised above the glass transition point for a certain period of time to allow the plastic to deform. In the third step, the stamp is separated from the polymer after cooling. The patterned polymer left on the substrate is used for further processing, such as dry etching or lift-off, or for use directly as a device component. Although the process is technically straightforward, there are several key issues that require special attention to make the process competitive as a nanofabrication technology as briefly summarized below. The first challenge for the nanoimprint lithography is the multilevel capability or the ability of exact

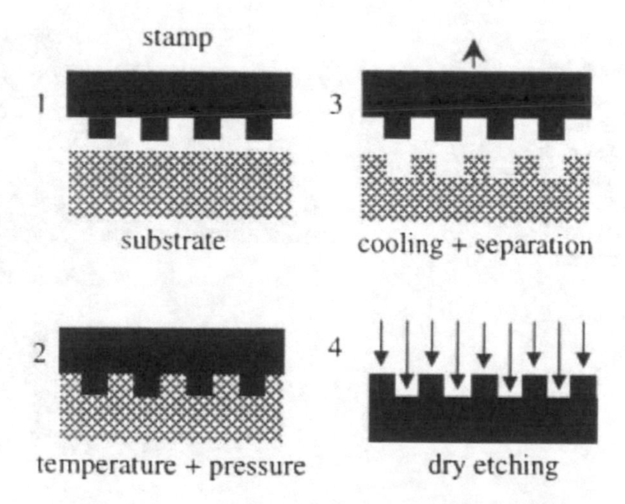

Fig. 11.9 Principal steps of a typical nanoimprint process. A stamp with the desired features is pressed on the polymer layer with the temperature raised above the glass transition point for a certain period of time to allow the plastic to deform. The stamp is separated from the polymer after cooling and the patterned polymer left on the substrate are used for further processing, such as drying etching or lift off, or for use directly as a device component. Reprinted with permission from © Design of Polymeric Platforms for Selective Biorecognition, 123–155 (2015)

alignment of multilayers. Various approaches have been explored to achieve exact alignment including use of commercially available stepper and aligner. Stamp size should be controlled, since large stamp size may introduce potential drawback, such as the parallelity of the substrate and stamp and thermal gradients in printing. The flow of the displaced polymer could set a limit to the feature density that imprinting stamps can achieve. Imprint of 50 nm features separated by 50 nm spaces within an area of $200 \times 200\ \mu m^2$ has been demonstrated. Sticking is another challenge to the nanoimprint lithography. Ideally, there should be no sticking at all between the polymer layers to be imprinted and the stamp. The choice of printing temperature, the viscoelastic properties of the polymer and the interfacial energy are among the key factors [21].

11.2.6 Dip-Pen Nanolithography

Dip-pen nanolithography is a direct-write method based upon an AFM and works under ambient conditions. Chemisorption is acted as a driving force for moving the molecules from the AFM tip to the substrate via the water filled capillary, when the tip is scanned across a substrate. Various nanostructures have been demonstrated and multicomponent nanostructures can readily be created. Figure 11.10 shows an example by DPN: fifteen nanometer dots spaced ~5 nm apart in the form of an "N" on an Au (111) substrate [22].

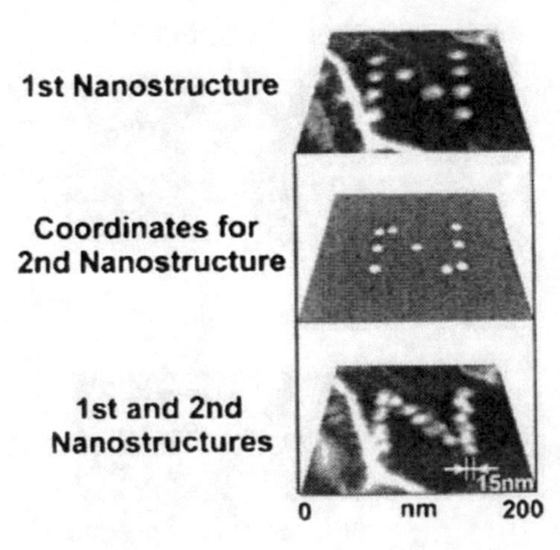

Fig. 11.10 SEM micrograph showing an example by dip-pen nanolithography: fifteen nanometer dots spaced ~5 nm apart in the form of an "N" on an Au (111) substrate. Reprinted with permission from © Imperial College Press, London (2004)

11.3 Etching

Etching is an important step in semiconductor device processing. Etching techniques are normally classified into two categories: wet and dry etching. For wet etch, the etchant is supplied from the liquid phase and can be carried out simply from a liquid container such as a beaker. For dry etch, etchant is supplied from the vapor phase. Common dry etching techniques include reactive ion etching (RIE), inductively coupled plasma (ICP)-RIE, chemical assisted ion beam etching (CAIBE), all of which require vacuum, plasma generation, ion optics etc. A widely used method is wet chemical etching. For example, dilute HF is used to etch a SiO_2 layer covering silicon. The HF reacts with SiO_2 and does not affect the photoresist or silicon. That is, this wet chemical etch is highly selective. However, the rate of etching is the same for any direction, lateral or vertical, so the etching is isotropic. Using an isotropic etching technique is acceptable only for relatively large structures. For nanosize structures, anisotropic etching with faster vertical etching is preferable [23].

Anisotropic etching generally exploits a physical process, or some combination of both physical and chemical methods. The best-known method of anisotropic etching is reactive-ion etching. Reactive-ion etching is based on the use of plasma reactions. This method works as follows. An appropriate etching gas, for example a chlorofluorocarbon, fills the chamber with the wafers. The pressure is typically reduced, so that a radiofrequency (RF) voltage can produce plasma. The wafer we want to etch is a cathode of this RF discharge, while the walls of the chamber are grounded and act as an anode. Figure 11.11 illustrates a principal scheme for the ion-etching method. The electric voltage heats the light electrons and they ionize gaseous molecules, creating positive ions and molecular fragments (so-called chemical radicals). Being accelerated in the electric field, the ions bombard the wafer normal to the surface. This normal incidence of bombarding ions contributes to the etching and makes the etching highly anisotropic. This process, unfortunately, is not selective. However,

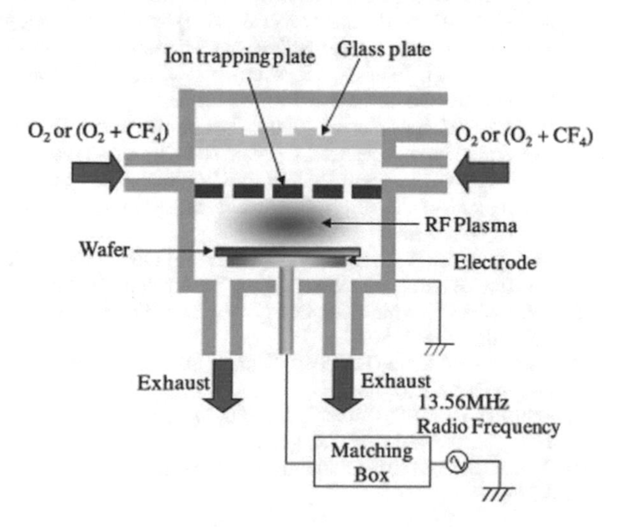

Fig. 11.11 Reactive-ion etching. Reprinted with permission from © Microelectronics Reliability, 52, 347–351 (2012)

the chemical radicals present in the chamber give rise to chemical etching, which, as we discussed, is selective. Now we see that the method combines both isotropic and anisotropic components and can give good results for etching on the nanoscale [24].

References

1. Mitin VV, Kochelap VA, Stroscio MA (2008) Introduction to nanoelectronics—science, nanotechnology, engineering, and applications. Cambridge University Press, New York
2. Suzuki K, Matsui S, Ochiai Y (2000) Sub-half-micron lithography for ULSIs. Cambridge University Press, Cambridge
3. Sun S, Mendes P, Critchley K (2006) Fabrication of gold micro- and nanostructures by photolithographic exposure of thiol-stabilized gold nanoparticles. Nano Lett 6:345–350
4. Xia Y, Rogers JA, Paul KE et al (1999) Unconventional methods for fabricating and patterning nanostructures. Chem Rev 99:1823–1848
5. Indykiewicz K, Macherzynski W, Paszkiewicz R (2011) The influence of contact mode on resolution in UV 400 lithography. In: International students and young scientists workshop, photonics and microsystems. https://doi.org/10.1109/stysw.2011.6155843
6. Singh M, Sun Y, Wang J (2012) Superconductivity in nanoscale systems. In: Superconductors—properties, technology, and applications. In Tech Publisher, Croatia
7. Barbillon G, Hamouda F, Bartenlian B (2014) Large surface nanostructuring by lithographic techniques for bioplasmonic applications. In: Manufacturing nanostructures. One Centre Press, UK
8. Wolfe DB, Christopher Love J, Whitesides GM (2004) Nanostructures replicated by polymer molding. In: Encyclopedia of nanoscience and nanotechnology. Marcel Dekker, New York
9. Zhang J, Shokouhi B, Cui B (2012) Tilted nanostructure fabrication by electron beam lithography. J Vac Sci Technol B 30:06F302–06F307
10. Peckerar M, Bass R, Rhee KW (2000) Sub-0.1 μ electron-beam lithography for nanostructure development. J Vac Sci Technol B Microelectron Nanometer Struct 18:3143–3149
11. Yang S, Wu Y (2018) EUV/soft x-ray interference lithography. In: Micro/nanolithography—a heuristic aspect on the enduring technology. Intech Open Science Publication, pp 83–100
12. Erdmanis M, Sievila P, Shah A (2014) Focused ion beam lithography for fabrication of suspended nanostructures on highly corrugated surfaces. Nanotechnol 25:335302–335309
13. Arshak K, Mihov M, Arshak A et al (2004) Novel dry-developed focused ion beam lithography scheme for nanostructure applications. Microelectron Eng 73:144–151
14. Thywissen JH, Johnson KS, Younkin R et al (1999) Nanofabrication using neutral atomic beams. J Vac Sci Technol B 15:2093–2100
15. Williams W, Saffman M (2006) Two-dimensional atomic lithography by submicrometer focusing of atomic beams. J Opt Soc Am 23:1161–1169
16. Li Cheung C, Camarero JA, Woods BW (2003) Fabrication of assembled virus nanostructures on templates of chemoselective linkers formed by scanning probe nanolithography. J Am Chem Soc 125:6848–6849
17. Notargiacomo A, Foglietti V, Cianci E et al (1999) Atomic force microscopy lithography as a nanodevice development technique. Nanotechnol 10:458–463
18. Qin D, Xia Y, Whitesides GM (2010) Soft lithography for micro- and nanoscale patterning. Nat Protoc 5:491–502
19. Kaufmann T, Ravoo BJ (2010) Stamps, inks and substrates: polymers in microcontact printing. Polym Chem 1:371–387
20. Schift H, David C, Gobrecht J (2000) Quantitative analysis of the molding of nanostructures. J Vac Sci Technol B 18:3564–3568
21. Barcelo S, Li Z (2016) Nanoimprint lithography for nanodevice fabrication. Nano Convergence 3:21–30

22. Cao G (2004) Nanostructures & nanomaterials synthesis, properties & applications. Imperial College Press, London
23. Marty F, Rousseau L, Saadany B (2005) Advanced etching of silicon based on deep reactive ion etching for silicon high aspect ratio microstructures and three-dimensional micro- and nanostructures. Microelectron J 36:673–677
24. Wang CG, Wu XZ, Di D et al (2016) Orientation-dependent nanostructure arrays based on anisotropic silicon wet-etching for repeatable surface-enhanced Raman scattering. Nanoscale 8:4672–4680

Chapter 12
Nanostructured Materials for Optical and Electronic Applications

Abstract Nanostructured materials in the area of optical and electronic applications are advancing more with more device development as solar panels, optoelectronic switches, batteries and sensors. This chapter particularly focuses on the optoelectronic applications. All electronic tools have one thing in common: an integrated circuit (IC) acting as their brain. Nano-electromechanical systems have evolved early years and creating sensors eyes and actuators arms at the same scale as the accompanying nanoelectronics. Recent developments in synthesis of nanomaterials with excellent electrical and mechanical properties have extended the boundaries of NEMS applications to include more advanced devices such as the non-volatile nano-electro-mechanical memory, where information is transferred and stored through a series of electrical and mechanical actions at the nanoscale (Venkateswara Rao and Kumar Yadav in Int J Eng Sci 4:6–9, 2015 [1]).

12.1 Applications of Nanostructured Materials in Solar Cells

The aim of device modeling is to develop a link between materials properties and the electrical device characteristics of a nano-structured solar cell. This is in contrast to materials modeling, where materials parameters like the optical absorption α (λ), the electrical mobilities μ_e, μ_h of electrons and holes, various relevant energy parameters, etc. are studied and theoretically modeled based on physical and chemical phenomena and interactions. In device modeling, we suppose that all materials parameters needed are already known, either from quantitative materials modeling or from experiments. The optical analysis of the cell already has been resulted in the knowledge of the optical generation of electron–hole pairs and excitons, both as a function of position and wavelength [2].

T. D. Thangadurai et al., *Nanostructured Materials*, Engineering Materials,
https://doi.org/10.1007/978-3-030-26145-0_12

12.1.1 Dye-Sensitized Nanostructured ZnO Electrodes for Solar Cell Applications

Dye-sensitized nanostructured solar cells (DNSCs) based on nanostructured metal oxide films have attracted more attention. They offer the prospect of low-cost photovoltaic energy conversion. Promising solar-to-electrical energy conversion efficiencies of more than 10% has been achieved and good progress has been made on long-term stability. The working mechanism of dye-sensitized solar cells differs completely from conventional p–n junction solar cells, but is, after more than 15 years of research, still not completely resolved. Research has largely focused on nanostructured TiO_2 (anatase) as the metal oxide to which the dye is bound. Good results have been obtained using other n-type metal oxides, such as ZnO, Nb_2O_5 and SnO_2. There will be focus on ZnO as a material for DNSC. ZnO is an attractive material for nanoscale optoelectronic devices, as it is a wide-band gap semiconductor with good carrier mobility and can be doped both n and p-type. The electron mobility is much higher in ZnO than in TiO_2, while the conduction band edge of both materials is located at approximately the same level. The material properties will be discussed in more detail in Section. A large range of fabrication procedures is available for ZnO nanostructures, such as sol–gel processes, chemical bath deposition, electrodeposition and vapor-phase processes. Different morphologies such as spherical particles, rods, wires and hollow tubes can be prepared with relative ease.

ZnO shows more flexibility in synthesis and morphology than TiO_2. The chemical stability of ZnO is, however, less than that of TiO_2, which can lead to problems in the dye adsorption procedure. The performance of dye-sensitized ZnO solar cells in terms of solar-to electrical energy conversion efficiencies is so far significantly lower than that of TiO_2, reaching currently about 4–5%. An analysis of the energetics and kinetics of ZnO-based DNSCs suggests that this is mainly due to the lesser degree of optimization in case of ZnO compared to TiO_2-based DNSCs. The main problem in dye-sensitized ZnO solar cells appears to be the troublesome dye adsorption process [3] (Fig. 12.1).

12.2 Photoconductive Oxide Nanowires as Nanoscale Optoelectronic Switches

Among all possible nano-devices, switches are critical for important applications like memory and logic. Electrical switching on nanometer and molecular level has been predominantly achieved through proper electrical gating configuration, as exemplified by nanotube transistors. It is possible to create highly sensitive electrical nanowire switches by exploring the photoconductivity of the individual semiconductor nanowires. It shows the conductance of the ZnO nanowire is extremely sensitive to ultraviolet light exposure. The light induced insulator-to-conductor transition allows us reversibly switching the nanowires from OFF to ON states, an “optical

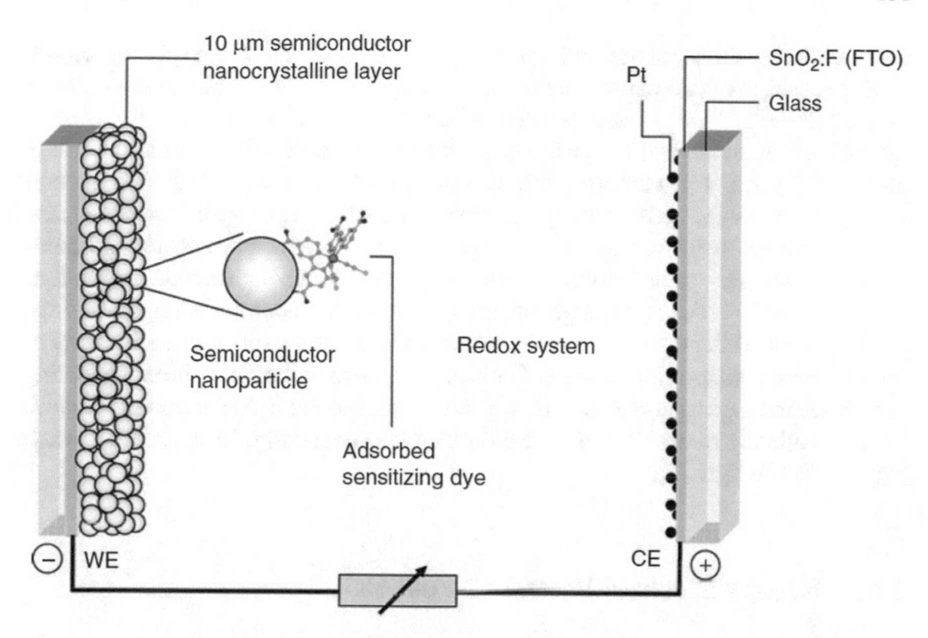

Fig. 12.1 Schematic illustration of the working principle of the DNSC. Incoming light excites the sensitizing dye, which induces an electron injection from the dye into the conduction band of the semiconductor nanoparticle. The electron is transported through the nanoporous structure to the back contact of the working electrode (WE). The electrons can then be utilized in work in an external circuit. The circuit is connected to the counter electrode (CE) where the redox system is effectively reduced with the help of a catalyst (nanoparticulate Pt in the case of I^-/I_3^-). The circuit is closed when the redox system reduces the sensitizing dye. Reprinted with permission from © Nanostructured Materials for Solar Energy Conversion, 227–254 (2006)

gating" phenomena as compared to commonly used electrical gating. Four-probe measurement of individual ZnO nanowires indicates that these nanowires are essentially insulating in dark with a resistivity above 3.5 M Ω cm^{-1}. When the nanowires were exposed to UV light with wavelengths below 400 nm, it was found that the nanowire resistivity instantly decreases by typically 4–6 orders of magnitude. In addition to the high sensitivity of the nanowire photoconductor, they also exhibit excellent wavelength selectivity.

The evolution of the photocurrent when a nanowire was exposed first to highly intense light at 532 nm (Nd:YAG, 2nd harmonics) for 200 s and then to UV light at 365 nm. There is no photoresponse at all to the green light while exposure to less intense UV-light shows the typical change of conductivity of 4 orders of magnitude. Measurements of the spectral response show that our ZnO nanowires indeed have a cut-off wavelength of 385 nm which is expected from the bandgap of ZnO [4].

It has been unambiguously established that oxygen chemisorption plays a profound role in enhancing photosensitivity of bulk or thin film ZnO. It is believed that a similar photoresponse mechanism could be applied to the nanowire system with additional consideration of high surface area of the nanowire, which could further enhance the sensitivity of the device. It was generally believed that the photoresponse of ZnO consists of two parts: a solid-state process where an electron and a hole are created ($h\nu \rightarrow h^+ + e^-$) and a two-step process involving oxygen species adsorbed on the surface. In the dark, oxygen molecules adsorb on the oxide surface as a negatively charged ion by capturing free electrons of the n-type oxide semiconductor ($O_2(g) + e^- \rightarrow O_2^-(ad)$) thereby creating a depletion layer with low conductivity near the nanowire surface. Upon exposure to UV-light, photogenerated holes migrate to the surface and discharge the negatively charged adsorbed oxygen ions ($h^+ + O_2^-(ad) \rightarrow O_2(g)$) through surface electron-hole recombination. Meantime, photo-generated electrons destruct the depletion layer, as a result, the conductivity of the nanowire increase significantly. The characteristics of the ZnO photoconducting nanowires indicates that they could be good candidates for optoelectronic switches, i.e. the insulating state as "OFF" in the dark and the conducting state as "ON" when exposed to UV-light [5].

12.3 Energy Storage, Batteries, Fuel Cells

One of the outstanding properties of nanotubes and other kinds of nanoparticles is their extremely large surface area, which can reach 2800 m^2/g in the case of carbon nanotubes. This means that ca. 2.5 g of nanotubes possess the same surface area as a soccer field (ca. $70 \times 100\ m^2$). This makes nano-structured materials ideal for applications in chemical processes relying on surface processes, e.g. catalysis, but also for the storage of energy, especially in the form of hydrogen. In fact, given the problems with fossil fuels, a new technology emerges to provide power supplies e.g. for transportation, buildings, and portable electrical devices, which is called hydrogen economy. It consists of the production of hydrogen from renewable or fossil energy sources e.g. by electrolysis of water, use it as a transportable energy carrier, and then convert it at the point of use to electricity e.g. by means of fuel cells. In each of these steps nanostructured materials, given their high surface to volume ratio, are envisioned to a play in future a most important role. However, nanostructured materials will also have a great impact on other technologies for the production, conversion, storage and application of energy (e.g. batteries, solar cells, etc.) [6].

The only data storage device was 5 1/4 floppy disk with a capacity of 110 kB, while the internal memory of his Commodore Pet computer was 32 kB. Today, use of CDs with a capacity of 700 MB, DVDs with 4 or more GB, and memory sticks with a capacity of 16 GB. In the case of memories, there are always two aspects which have to be kept in mind: The data have to be stored in a medium by some kind of effect, but that also have to be written and read within a reasonable time, which at least in the case of reading is defined by real time reproduction of music, films etc. As in the case of semiconductor devices, higher storage densities have been achieved by reducing the sizes of a bit, and similar to semiconductor devices, the technologies are currently approaching limits by reaching the nanometer region with

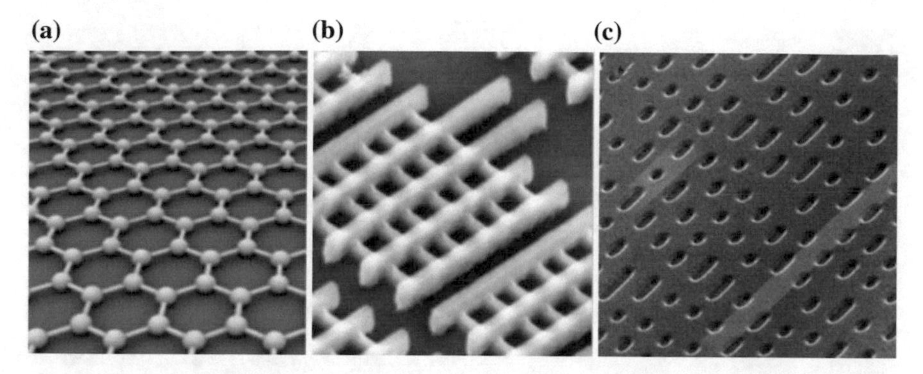

Fig. 12.2 **a** Carbon-carbon bond length in graphene: 0.142 nm; **b** length of a transistor gate in the latest CPU technology from Intel: 22 nm; **c** size of a single bit of information on a DVD disk: 400 nm

the consequences emphasized throughout this contribution. In the case of technique using ferromagnets as storage medium (such as hard disks), for example, a lower limit of the size of a bit is given by the superparamagnetic effect. For compact discs and their successors which rely on optical readout, the wavelength of the lasers used has been reduced from infrared (CD) over red (DVD) to blue (Blue-ray discs). The pit size for the latter is 160×320 nm^2. However, future improvement will very probably require completely new approaches. One possibility is the use of optical near-field techniques for which the Rayleigh criterion is not applicable, or the use of three dimensional approaches, e.g. holographic techniques. In the former case, the reading head must be situated a few nanometers only from the storage medium. A large research effort is presently devoted to the so-called phase change materials, which exhibit drastic differences of their electrical (resistivity) and optical properties (refractive index) properties, depending on their structure (crystalline or amorphous). As a change of the structure can be easily achieved in very short times by heating with a laser beam either to the crystallization or to the melting temperature, data can be written optically, while the readout can take place either electrically or optically [7] (Fig. 12.2).

12.4 Nanostructured Semiconductor Materials for Optoelectronic Applications

Semiconductor nanostructures allow the control of basic material parameters only by changing geometric factors without changing the material composition. To understand the strong impact of geometric factors in the nanometer regime one has to briefly review the bulk properties of semiconductors. Most of the semiconductors used in electronics and optoelectronics form diamond, like silicon (Si), or zinc blende lattices, like gallium arsenide (GaAs) or indium phosphide (InP) as shown in Fig. 12.3.

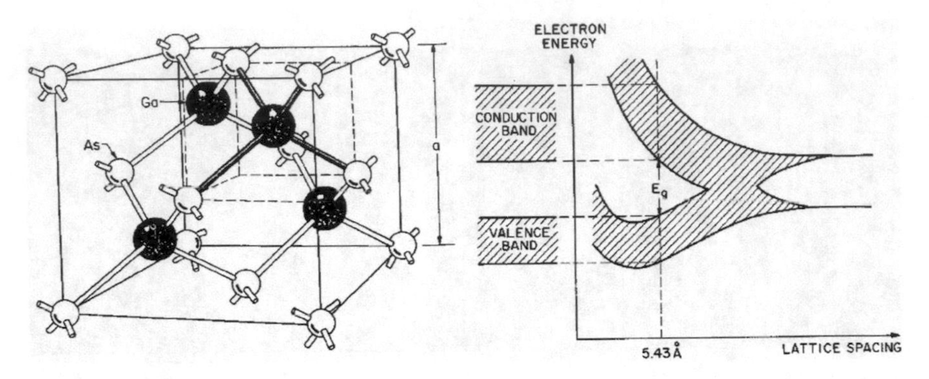

Fig. 12.3 Zinc blende lattice of, e.g., GaAs or InP (left). Electronic band structure formation as a function of the lattice spacing for a diamond type lattice (right). Reprinted with permission from © Nanostructured Materials for Advanced Technological Applications, Springer (2008)

During the formation of a crystal from N isolated atoms, e.g., from a liquid or gas phase, the atomic spacing decreases and the electronic valence states of each atom overlap each other forming bands with N states. However, the energetic distances are too small to be distinguishable. In case of the formation of a periodic lattice the electronic wave functions can interfere with the periodic potential fluctuation of the crystal and build a band structure with bands and band gap regions, where no electron can penetrate into the crystal [8].

Nanostructured materials lead to new directions in the fabrication of complex shape microparts for microdevices. For instance, it was demonstrated that hot embossing on UFG AA1050 processed via ECAP shows a good potential for application in fabrication of microheat exchangers with geometrical features smaller than 10 μm and high thermal conductivity. Hot embossing on this material provided a very smooth embossed surface with fully transferred pattern and low failure rate of the mold, while hot embossing on the coarse-grained AA1050 resulted in a much rougher surface with shear bands. Figure 12.4 shows microforming process used to fabricate a microturbine from ECAP-processed UFG pure Al at ambient temperature. Subsequent examination demonstrated that UFG pure aluminum gives much higher

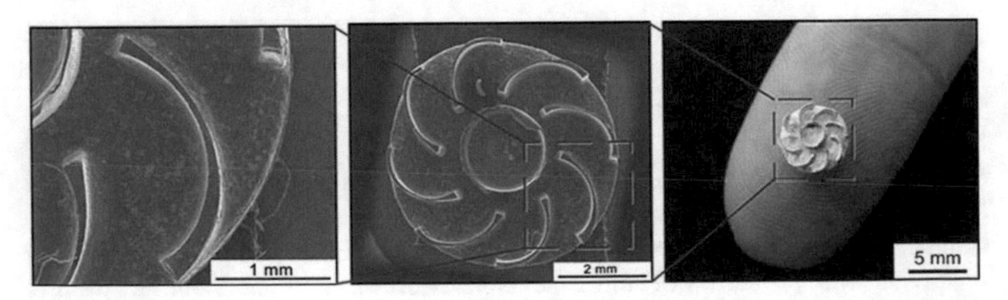

Fig. 12.4 Microturbine of UFG pure aluminum formed at ambient temperature. Reprinted with permission from © Advanced Engineering Materials, 17, 1022–1033 (2016)

strength and more uniform mechanical properties by comparison with conventional coarse-grained pure aluminum. Downscaling of the SPD processing techniques along with combination with another processing method can dramatically ease fabrication of microparts for microdevices. A nice example was presented in, where a miniaturized ECAP die with a millimeter-scale channel was developed to produce Al wires with grain size of 1–2 μm in a single ECAP step showing enhanced strength. This ECAP preprocessing can be combined with a final extrusion step where a desired axisymmetric profile can be easily imparted to the final product. An example of a possible product is a long bar with a cogwheel profile that can be chopped easily into MEMS gears [9].

12.5 Carbon-Based Sensors and Electronics

The semiconductor industry has been able to improve the performance of electronic systems for more than four decades by downscaling silicon-based devices but this approach will soon encounter its physical and technical limits. This fact, together with increasing requirements for performance, functionality, cost, and portability have been driven the microelectronics industry towards the nano world and the search for alternative materials to replace silicon. Carbon nanomaterials such as one-dimensional (1D) carbon nanotubes and two-dimensional (2D) graphene have emerged as promising options due to their superior electrical properties which allow for fabrication of faster and more power-efficient electronics. At the same time their high surface to volume ratio combined with their excellent mechanical properties has rendered them a robust and highly sensitive building block for nanosensors. In 2004, it was shown for the first time that a single sheet of carbon atoms packed in a honeycomb crystal lattice can be isolated from graphite and is stable at room temperature. The new nanomaterial, which is called graphene, allows electrons to move at an extraordinarily high speed. This property, together with its intrinsic nature of being one-atom-thick, can be exploited to fabricate field-effect transistors that are faster and smaller [10] (Fig. 12.5).

When a layer of graphene is rolled into a tube, a single-walled carbon nanotube (SWNT) is formed. Consequently, SWNTs inherit the attractive electronic properties of graphene but their cylindrical structure makes them a more readily available option for forming the channel in field-effect transistors. Such transistors possess an electron mobility superior to their silicon-based counterpart and allow for larger current densities while dissipating the heat generated from their operation more efficiently. During the last decade, carbon nanotube-based devices have advanced beyond single transistors to include more complex systems such as logic gates and radio-frequency components [11] (Figs. 12.6 and 12.7).

Recent advances in nanofabrication techniques have provided the opportunity to use single molecules, or a tiny assembly of them, as the main building blocks of an electronic circuit. This, combined with the developed tools of molecular synthesis

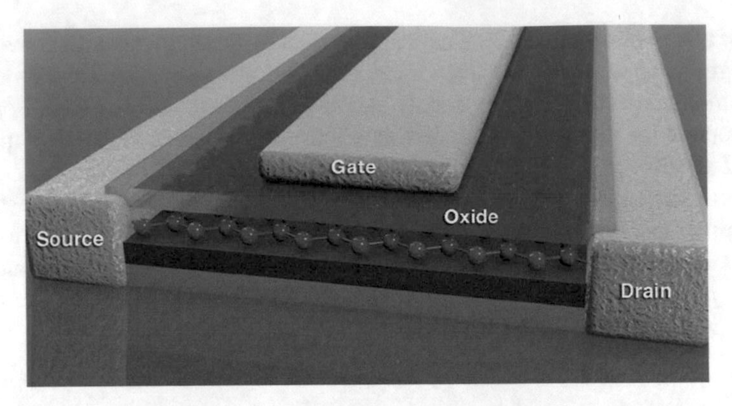

Fig. 12.5 The layer of graphene acts as the conducting channel in a field-effect transistor

Fig. 12.6 The artistic expression of an integrated circuit based on individual carbon nanotubes. Reprinted with permission from © Cees Dekker, TU Delft/Tremani

to engineer basic properties of molecules, has enabled the realization of novel functionalities beyond the scope of traditional solid state devices. A modern memory device, in its most common implementation, stores each bit of data by charging up a tiny capacitor. The continuous downscaling of electronic circuits, in this context, translates to storing less charge in a smaller capacitor. Ultimately, as memory device dimensions approach the nanometer range, the capacitor can be replaced by a single

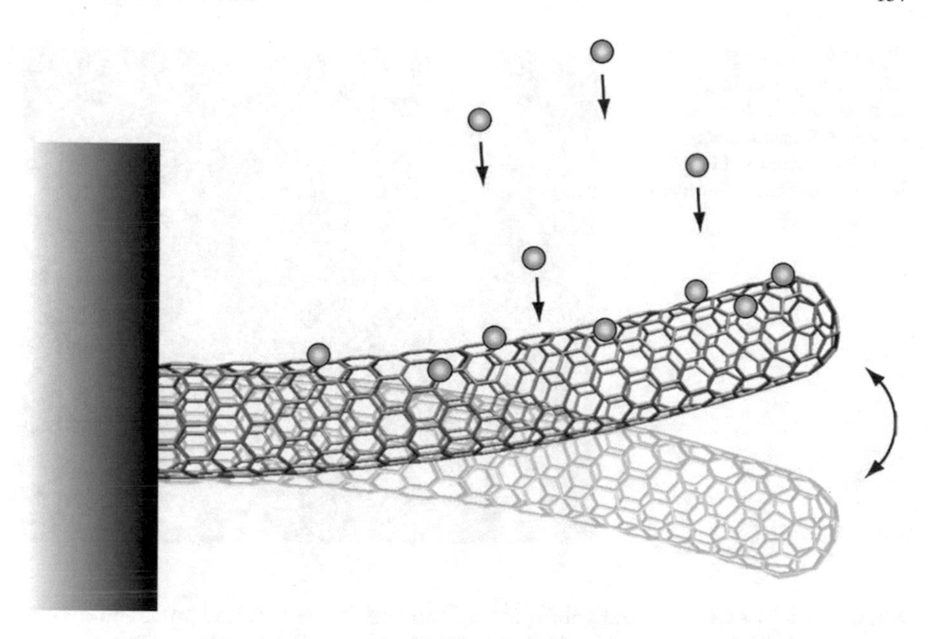

Fig. 12.7 The additional gold atom that adsorbs on the surface of a vibrating carbon nanotube would change its resonance frequency which is further detected. Reprinted with permission from © Zettl Research Group, Lawrence Berkeley National Laboratory and University of California at Berkeley

organic molecule such as Ferrocene, whose oxidation state can be altered by moving an electron into or out of the molecule [12]. In contrast to common transistors, where the switching action requires thousands of electrons, a single electron transistor needs only one electron to change from the insulating to the conducting state.

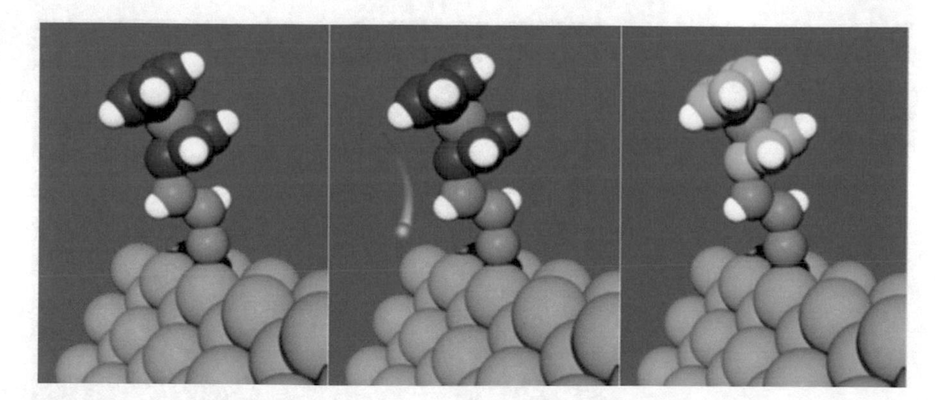

Fig. 12.8 A neutral Ferrocene molecule is attached to a nanoelectrode representing a "0" state. An electron tunnels to the nanoelectrode by the application of an external electrical field. The positively charged Ferrocene molecule represents a "1" state

Fig. 12.9 The single electron transistor in a surface acoustic wave echo chamber. Reprinted with permission from © Philip Krantz, Chalmers University of Technology

Such transistors can potentially deliver very high device density and power efficiency with remarkable operational speed. In order to implement single electron transistors, extremely small metallic islands with sub-100 nm dimensions have to be fabricated. These islands, which are referred to as quantum dots, can be fabricated by employing processes made available by the advances in nanotechnology [13] (Figs. 12.8 and 12.9).

While conventional electronic devices rely on the transport of electrical charge carriers, the emerging technology of spintronics employs the spin of electrons to encode and transfer information. Spintronics has the potential to deliver nanoscale memory and logic devices which process information faster, consume less power,

Fig. 12.10 The close-up look at a hard disk drive improved with the Giant Magneto-Resistance technology

and store more data in less space. The extension of the hard disk capacities to the gigabyte and the terabyte ranges was the main achievement of spintronics by taking advantage of Giant Magneto-Resistance (GMR) and Tunnel Magneto-Resistance (TMR) effects which are effective only at the nano scale [14] (Fig. 12.10).

References

1. Venkateswara Rao A, Kumar Yadav S (2015) An introduction to nano electro mechanical systems. Int J Eng Sci 4:6–9
2. Afshar EN, Xosrovashvili G, Rouhi R et al (2015) Review on the application of nanostructure materials in solar cells. Mod Phys Lett B 29:1550118–1550124
3. Giannouli M (2013) Nanostructured ZnO, TiO_2, and composite ZnO/TiO_2 films for application in dye-sensitized solar cells. Int J Photoenergy 612095:1–8
4. Cammi D, Ronning C (2014) Persistent photoconductivity in ZnO nanowires in different atmospheres. Adv Cond Matter Phys 184120:1–5
5. Kind H, Yan H, Messer B et al (2002) Nanowire ultraviolet photodetectors and optical switches'. Adv Mater 14:158–160
6. Schneemann A, White JL, Young Kang S et al (2018) Nanostructured metal hydrides for hydrogen storage. Chem Rev 118:10775–10839
7. Shih CC, Lee WY, Chen WC (2016) Nanostructured materials for non-volatile organic transistor memory applications. Mater Horiz 3:294–308
8. Peter Reithmaier J, Petkov P, Kulisch W et al (2008) Nanostructured materials for advanced technological applications. In: Proceedings of the NATO advanced study institute on nanostructured materials for advanced technological applications, Sozopol. Springer, Netherlands
9. Qiao XG, Gao N, Moktadir Z et al (2010) Fabrication of MEMS components using ultra-fine grained aluminium. J Micromech Microeng 20:045029–045041
10. Mina AN, Phillips AH (2013) Graphene transistor. JASR 9:1854–1874
11. Bachtold A, Hadley P, Nakanishi T et al (2001) Logic circuits with carbon nanotube transistors. Science 294:1317–1319
12. Credi A, Semeraro M, Silvi S et al (2011) Redox control of molecular motion in switchable artificial nanoscale devices. Antioxid Redox Signal 14:1119–1165
13. Hu P, Zhang J, Li L et al (2010) Carbon nanostructure-based field-effect transistors for label-free chemical/biological sensors. Sensors 10:5133–5159
14. Gautam A (2012) Spintronics—a new hope for the digital world. IJSR 2:1

Chapter 13
Nanostructured Materials for Bioapplications

Abstract Nanostructured materials development has high impact on biological, biomedical and clinical applications. The nanostructural materials applied for biological system can be an interface with high biodegradability and minimum toxicity. That concern must to be resolve with design and synthesis of nanoscale structures. This chapter describes more on biomedical application and their design techniques.

Recent improvements in nanostructured materials and nanotechnology will have profound impact in many areas such as energy technologies and biomedical applications. In the biomedical applications, traditional materials science and engineering face new challenges in the synthesis and microstructure development since the requirements for general materials must be based on special medical needs. The most fascinating development in nano-biomedicine is to be found in biomedical diagnosis and treatment, and involves the direct use of nanomaterials within a biological system. The in vivo imaging by fluorescent nanoparticles such as quantum dots is progressing rapidly; and cell targeting via surface functionalized nanoparticles is undergoing animal tests and should be available within a few years. The nanoscaled systems with surface functionalized groups that is able to conjugate with a variety of biological molecules including DNA, RNA, and viruses [1].

Adaptation of nanostructured materials into biomedical devices and systems has been of great interest. Through the modification of existing nanostructured materials one can control and tailor the properties of such materials in a predictable manner, and impart them with biological properties and functionalities to better suit their integration with biomedical systems. These modified nanostructured materials can bring new and unique capabilities to a variety of biomedical applications ranging from implant engineering and modulated drug delivery, to clinical biosensors and diagnostics [2]. Silicon Quantum dots (SiQDs) and Silicon Nanowires (SiNWs) are attractive low-dimensional silicon nanostructures that have found important applications in biology, particularly fluorescence bioimaging and ultrasensitive biosensing. SiQDs-based fluorescent biological probes Fluorescent biological probes are powerful tools for biological and biomedical studies.

A high-performance fluorescent cellular probe should be water-dispersible, highly fluorescent, anti-photobleaching and biocompatible. Organic dyes and fluorescent

T. D. Thangadurai et al., *Nanostructured Materials*, Engineering Materials,
https://doi.org/10.1007/978-3-030-26145-0_13

proteins have been widely used in biological studies. Notwithstanding, these dyes and proteins usually suffer from severe photobleaching that hampers long-term imaging in vitro or in vivo. In the past, the semiconductor II–VI QDS have been developed as high-performance fluorescent biological probes because of advantages such as size-tunable emission color, strong fluorescence and high resistance to photobleaching. However, the potential toxicity problem of the II–VI QDs associated with release of heavy metal ions (e.g. Cd ions) has not yet been fully addressed, which limits their widespread biological and medical applications. Consequently, novel fluorescent probes with robust photostability, strong fluorescence and favorable biocompatibility are still urgently required to satisfy various requirements of biological studies [3].

13.1 Nanostructured Ti and Ti Alloys for Biomedical Engineering

Pure Ti possesses the highest biocompatibility with living organisms, but it has limited use in medicine due to its low strength. High-strength nanostructured pure Ti processed via SPD opens new avenues and concepts in medical device technology. First, expensive and toxic alloying elements are absent in the material. Thus, the implants made therefrom have better biocompatibility. Second, implant size can be noticeably reduced decreasing the level of surgical intervention into human body. Third, the material demonstrates improvement of biological reaction on its surface. A new SPD-based technology has been recently developed for the fabrication of nanostructured Ti for dental implants. The processing route consists of Equal channel angular extrusion ECAP-C leading to grain refinement and secondary processing swaging provides shaping and additional strengthening. Long rods having lengths up to 3 m, diameters of 4–8 mm, and accuracy grade h8 suitable for automation of implant machining can be produced using this method. Mechanical properties of this nanostructured Ti are even higher compared to those of conventional high-strength Ti alloys (such as Ti–6Al–4 V) used in biomedical engineering [4] (Fig. 13.1).

13.2 Nanostructured Materials for Biosensors

Biosensors based on nanostructured metal oxides gained much attention in the field of health care for the management of various important analytes in a biological system. The unique properties of nanostructured metal oxides offer excellent prospects for interfacing biological recognition events with electronic signal transduction and for designing a new generation of bioelectronics devices. A large number of nanostructured metal oxides such as cerium oxide (CeO_2), iron oxide (Fe_3O_4), magnesium oxide (MnO_2), niobium oxide (Nb_2O_5), nickel oxide (NiO), praseodymium oxide (Pr_2O_6), tin oxide (SnO_2), titanium oxide (TiO_2), zinc oxide (ZnO) and zirconium

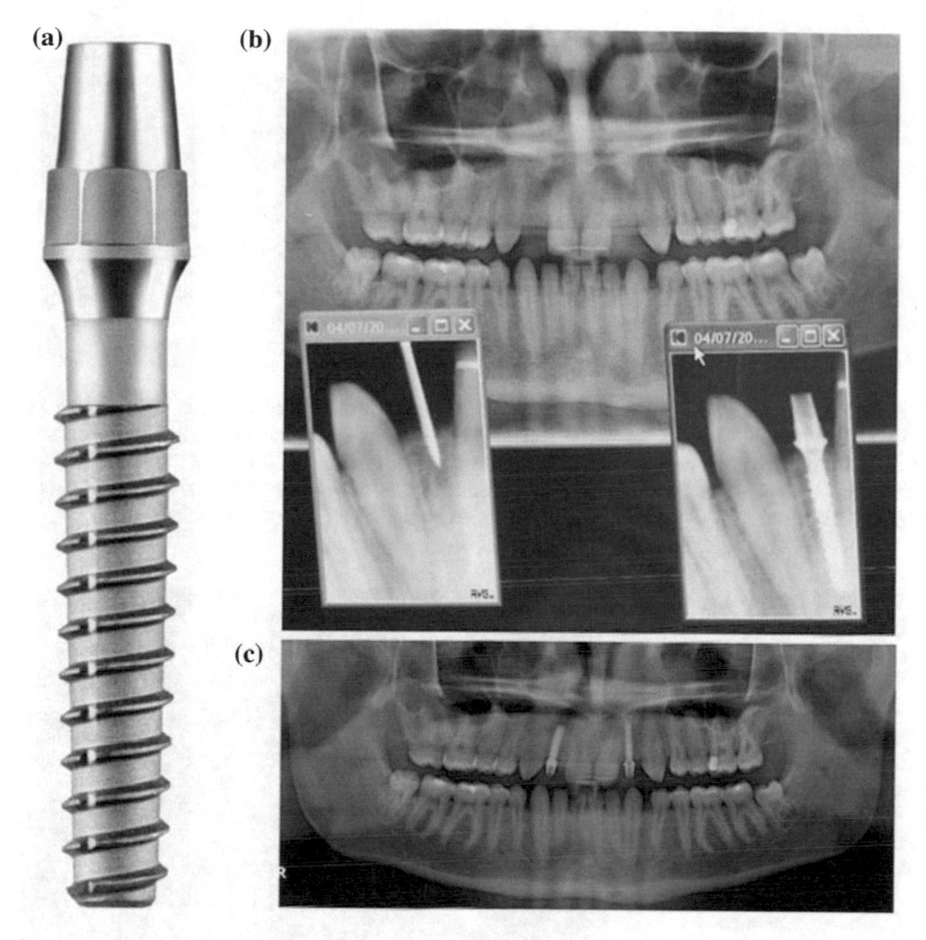

Fig. 13.1 **a** Dental implant from nanostructured Ti, **b** X-ray photographs after surgery and **c** control photograph after incorporation of dental implants into human jaw

oxide (ZrO_2) have been used for their application in electrochemical biosensors [5] (Figs. 13.2 and 13.3).

13.3 Nanobiotechnology

Nanobiotechnology is one of the most important applications of nanotechnology. There are several reasons driving this development: (i) many biomolecules and bio-entities are of nanometer size. Even cells, which are in the micrometer range, consist of nanosized component. Thus, the nanometer range is best suited to study the interactions between bio-entities and non-biological materials, (ii) Adhesion, growth, proliferation and viability of cells strongly depend not only on the chemical nature

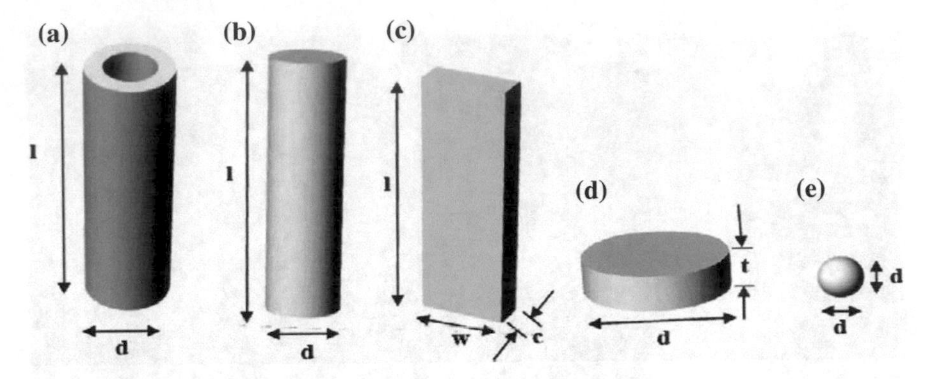

Fig. 13.2 The various form of nanostructures with typical dimensions. **a** Nanotube, l: length (greater than 1000 nm), d: diameter (less than 100 nm); **b** nanowire, l: length (greater than 1000 nm), d: diameter (less than 100 nm); **c** nanobelt, l: length (greater than 1000 nm), w: width (less than 500 nm), c: depth (less than 100 nm); **d** nanodiskette, t: thickness (less than 100 nm), d: diameter (generally between 500–1000 nm); **e** nanoparticles, d: diameter (order of few nanometers)

Fig. 13.3 The true example of nanotechnology: an array of individually addressable vertically-aligned carbon nanofibers for sensing applications at the nanoscale. For comparison, a single human hair is 1000 times thicker than any of the nanofibers in the image. Reprinted with permission from © Farzan Alavian Ghavanini, Chalmers University of Technology

of the substrate but also on its nanostructure, (iii) Biosensors will play an ever increasing role in modern medicine. For the development of such sensors, an understanding of the interactions between biomolecules and surfaces on the nanoscale is imperative. On the other hand, devices such as labs-on-a-chip require the sensors themselves to be of micro- or even nanometer size, (iv) Nanoparticles and other nanostructured materials are envisioned to play an important role in drug delivery, i.e. the transport of drugs and other medical substances to the very place of use in the body, and also in medical imaging. (v) Nature herself very often uses nanostructured approaches to realize astonishing material properties. The famous lotus effect is but one example. The nacre of abalone shells is formed by a laminated structure consisting of layers of inorganic material ($CaCO_3$) of about 0.5 μm thickness which are separated by very thin organic layers of less than 10 nm. Such a structure is simultaneously hard, strong, and tough [6].

13.4 Gene Therapy

Gene carriers holds excessive potential for the prevention or treatment of certain diseases and genetic disorders by delivering therapeutic nucleic acids into the defective cells, thereby adjusting and controlling the corresponding cellular processes and responses. The delivery of therapeutic nucleic acids into cells is one of the major hurdles for the successful gene therapy. The therapeutic nucleic acids (pDNA or siRNA) need to be shuttled and successfully transferred into the defective cells by gene carriers. One way to deal with this problem is through viral based gene carriers. Although viral gene carriers are efficient, the clinical application based on this approach is restricted by several safety concerns, including immunogenicity, carcinogenicity, immune response, and virus replication. Some of these shortcomings may be overcome by applying non-viral carriers. Polymeric gene carriers have exhibited some unique advantages, such as safety, physiological stability and suitable for large-scale production. However, the transfection efficiency of the non-viral carriers is usually lower than that of the viral carriers. Generally, the complexes that the non-viral carriers form with nucleic acids for gene therapy can be classified into four categories: polyplex, lipoplex, micelleplex and others [7].

13.5 Bioimaging

Bioimaging is a dominant technique that can directly observe normal and abnormal biological processes in individual patients. Many bioimaging modalities have been developed, tested and utilized in preclinical and clinical applications in the past two decades. However, the applications of this technique are often hampered by the poor sensitivity, specificity, and targeting ability of the available and suitable bioimaging probes. The typical polymeric nanostructured bioimaging probes are nanoassemblies consisting of a bioimaging core and a polymer coating as shell. The polymer coating not only protects the loaded probes from the environment, but also improves the pharmacokinetics and bio-distribution of the probes, thus significantly amplifying the diagnostic imaging signals. The developed modalities include contrast agents for magnetic resonance imaging (MRI), X-ray computed tomography (CT), fluorescence imaging (FI), Single-photo emission computed tomography (SPECT) and positron emission tomography (PET) [8].

13.6 Tissue Engineering and Regenerative Medicine

Nanotechnology and nanoengineering are effective means in the design, preparation, characterization and applications of nanoscale devices, which consist of functional organizations with at least one dimension in the range from several to hundreds nanometers. Recently, the synergy between nanoscience and tissue engineering has led to great developments in biomedical research as well as clinical practices, including the realms of bone and cartilage regeneration, blood vessel tissue engineering, wound dressing, and so on [9].

13.7 Bone Implant

The implants and scaffolds for bone repair and regeneration are of considerable interest for biomedical applications of nanostructured materials, as natural bone exhibits certain hierarchical structures of nanometer dimensions within bone matrices. In fact, compared to respective micro-sized counterparts most nanoscale bone implant materials, such as ceramics like alumina, titania, and hydroxyapatite (HA), polymers (PLGA and polyurethane), and metals (Ti, Ti6A14V, and CoCrMo alloy), have been shown to enhance bone cell responses and functions that include cellular adhesion, proliferation, synthesis of alkaline phosphatase, and calcium deposition. In spite of the enhancement of cellular responses, poor mechanical properties of the nano-sized implant materials limit their further applications for bone repair. Poor mechanical properties restricts clinical use of nano-HA as load-bearing implants, and limit their applications to small, unloaded and lightly-loaded implants, powders, coatings, composites, and porous scaffolds for tissue engineering. Furthermore, the bending modulus of nanophase alumina was 1.8 times greater than that of natural human femur bone, even though the nanophase ceramic has closer bending properties to native bone than conventional ceramics. Mismatch in elastic modulus between alumina and the surrounding bone often causes a stress shield effect in bone tissue, resulting in bone resorption and loosening of implants. To compensate for these weak features, modifications of nanostructured bone implant materials are needed to improve their mechanical properties. Appropriate modifications can also further reinforce the interface between bone cells and the implant matrix. Different methods have been proposed to modify existing nanostructured materials for bone implants [10].

13.8 Modulated Drug Delivery System

In nanoparticle systems, including solid nanoparticles, polymeric nanoparticles, and polymeric self-assemblies, have attracted increasing attention for use as potential drug delivery systems. The advantages of using nanoparticle systems for drug delivery result from their two basic properties. First, nanoparticulate, due to their small size, can penetrate small capillaries and be taken up by cells, which allows for efficient drug accumulation at target sites in the body. Second, the use of biodegradable materials for nanoparticulate preparation allows for sustained drug release within the target over a period of days or even weeks after administration. Furthermore, new functions arising from nano-sizing, such as improved solubility, target ability and adhesion to tissues can be conferred to nano-delivery systems and also provided the possibility to covert poorly soluble, poorly absorbable, and predisposed biologically active substances into promising drugs. In spite of these advantages, however, some technical problems limit clinical applications of these new nanoparticle drug delivery systems. Some of these limitations include relatively short blood circulation time, low accessibility to physiological barriers of target sites blood-brain barrier, lymphatic barrier, etc., and low efficiency of gene transfection. Suitable and effective modifications of these nanoparticle materials are needed to overcome these technique obstacles [11].

13.9 DNA Biosensor

Biosensors for DNA are based on the process of hybridization, the matching of one strand of DNA with its complementary. These biosensors can be used for recognition and quantization of target DNA in clinical samples. DNA analysis is widely considered to be the most recent and promising use of biosensors for clinical applications, especially for genomic sequencing, mutation detection, and pathogen identification of inherited diseases. Similar to antibody and antigen biosensors, the introduction of nanostructured materials with appropriate modifications into DNA biosensors greatly improves properties including sensitivity and broadening of analytes. Bimolecular grafting, or coupling, is one of the main ways to modify nano-structured materials for DNA biosensors. DNA-functionalized gold nanoparticles have been used in quartz crystal microbalance (QCM) to enhance DNA sensor sensitivity. In order to avoid the high negative charge density of DNA, neutral peptide nucleic acid (PNA), DNA analogues with the entire sugar-phosphate backbone replaced by a polyamide backbone, have been prepared to modify gold nanocrystals for DNA sensing. The PNA complexes could offer nanomaterials with two distinct advantages: (1) greater stability than DNA duplexes and (2) greater mismatch sensitivity. The latter feature enables PNA-modified nanoparticles to act as highly selective nanoscale sensors. Furthermore, when coupled with a substantial change in colloidal stability upon DNA hybridization, PNA-modified nanoparticles can be used to develop novel colorimetric DNA assays that detect the presence of single base imperfections within minutes [12].

13.10 Glucose Biosensor

Blood glucose is a clinically important analyte for diabetic health care. In most glucose sensors, such as oxygen, hydrogen peroxide, and mediator detectors are based on electrochemical amperometry and use a glucose oxidase (GOD) as the recognition element. The main concern when designing enzyme based amperometric biosensors is how to effectively transfer electrons to the electrodes. Carbon nanotubes (CNT) have successfully been applied to glucose biosensor platforms to improve the electronic properties because of their ability to promote electron-transfer reactions with enzymes. However, to further enhance the function of glucose biosensors and maintain the biological activity of GOD in the system, modifications of the CNT are necessary. It was reported that Pt nanoparticles are very effective as matrix of enzyme sensors because of their biocompatibility, large surface, and good electrocatalytic activity with hydrogen peroxide. In order to overcome the difficulty of the deposition of Pt nanoparticles onto CNT, due to the very high hydrophobicity of CNTs, further surface modifications and sensitization activations of CNT surfaces were needed. Nafion, a pefluorosulfonated and negatively charged polymer, was found to improve depositions of Pt nanoparticles on CNTs, due mainly to the charged interactions. In addition to having a higher sensitivity to glucose than GOD electrode modified with Pt nanoparticles or CNT alone, Pt nanoparticle-CNT GOD electrodes exhibited other superior properties, such as excellent electrocatalytic activity, large determination range, short response time, large current density, and high stability [13].

13.11 Therapies

Multimode imaging is of paramount importance for the identification and diagnosis of the disease, but the treatment of the disease, especially targeted therapy, is more important in clinical medicine. Some nanostructures containing UCNPs as therapeutic agent could be used for therapy of diseases, especially for tumor therapy. PTT, PDT and chemotherapy are the most common therapeutic methods based on upconversion nanostructures [14].

13.11.1 Photodynamic Therapy

Compared to surgery, chemotherapy and radiotherapy, PDT is a noninvasive cancer treatment. PDT is involved, in which PDT drugs (photosensitizers) are activated by UV, visible light or NIR to generate cytotoxic reactive oxygen species (ROS) for killing the target cells. The organic molecules, such as methylene blue (MB), zinc (II) phthalocyanine (ZnPc), Chlorine6 (Ce6) and so on, and semiconductor nanomaterials are the most common photosensitizers for PDT. Because UV and visible light with low penetration ability limit the development of PDT, utilizing UV and visible light converted by UCNPs from NIR can overcome the limitation of low penetration ability. Semiconductor nanomaterials, like ZnO or TiO_2, as photosensitizer, could be deposited on the surface of UCNPs to construct a core–shell nanostructure for PDT. The UCNPs@TiO_2 nano constructures are used for effective PDT. The absorption band of TiO_2 shell matched well the UV emission from NIR converted by UCNPs.

So TiO_2 shell can be activated by NIR to generate and release ROS, which could suppress tumor growth efficiently. In order to overcome the tumor hypoxia environment to achieve effective treatment, many ideas have been proposed. An intelligent upconversion nano theranostic system (TPZ-UC/PS) has been designed. In this system, PS could be activated by UV emission from NIR converted by UCNPs to produce ROS, so that the environment was hypoxia and TPZ was highly cytotoxic under hypoxia. The synergistic treatment of ROS and TPZ was effective for tumors [15].

13.11.2 Chemotherapy

Although chemotherapy is invasive, it is still an effective method for disease treatment. The chemotherapy based on UCNPs includes two aspects: imaging-guided chemotherapy and photo trigger-induced chemotherapy.

Imaging-guided chemotherapy could indirectly observe and monitor the extent of drug release by using the UCL of UCNPs. SiO_2, mesoporous SiO_2 or polymer coated UCNPs with porous or hollow structure are good drug and siRNA carrier for chemotherapy. Because of the acidic environment of tumor, some cases of drug release are controlled by pH. For example, Lin's group reported sub-10 nm $BaGdF_5$:Yb/Tm UCNPs as drug carrier. After modified by gelatin, the UCNPs could be conjugated with DOX via covalent interactions. The drug release of DOX in acidic environment (tumor) was faster than that in neutral environment (normal tissue), due to the cleavage of hydrazone bonds between DOX and UCNPs in acidic environment. The pH trigger-guide drug release combing UCL/MR/CT multimodal imaging has a great potential for simultaneous diagnosis and therapy of diseases [16].

13.11.3 Photothermal Therapy

PTT is relative noninvasive treatment for cancer diseases, whose core is photothermal agent. The photothermal agents could absorb light and then convert it into heat to cause thermal damage for cancer cells. PTT could be achieved via combining UCNPs with nanoparticles with photothermal function. Recently, Au and Ag with surface plasmon resonance absorption are widely used as photothermal agents for PTT. Therefore, combining Au or Ag nanoparticles with UCNPs is an effective agent for PTT. For example, the synthesis of a novel class of multifunctional nanoparticles, UCNPs, Fe_3O_4 nanoparticles, and Au nanoparticles via layer-by-layer self-assembly to be used for multifunctional bioimaging and PTT. Graphene oxide (GO) and Cu_xS were also used as photothermal agents for tumor therapy. The synthesis of GO covalently grafted UCNPs and then loading ZnPc on GO, which could act as a theranostic platform for UCL bioimaging and PTT/PDT of cancer [17].

References

1. Shi D, Gu H (2008) Nanostructured materials for biomedical applications. J Nanomater 529890:2
2. Xu T, Zhang N, Nichols HL et al (2007) Modification of nanostructured materials for biomedical applications. Mater Sci Eng C 27:579–594
3. He Y, Fan C, Lee ST (2010) Silicon nanostructures for bioapplications. Nano Today 5:282–295
4. Morais LS, Serra GG, Muller CA et al (2007) Titanium alloy mini-implants for orthodontic anchorage: immediate loading and metal ion release. Acta Biomater 3:331–339
5. Castillo J, Andersen KB, Svendsen WE (2011) Self-assembled peptide nanostructures for biomedical applications: advantages and challenges. In: Pignatello R (ed) Biomaterials science and engineering. InTech Publications, London
6. Fakruddin M, Hossain Z, Afroz H (2012) Prospects and applications of nanobiotechnology: a medical perspective. J Nanobiotechnol 10:31–39
7. Cavazzana-Calvo M, Hacein-Bey S, De Saint Basile G et al (2000) Gene therapy of human severe combined immunodeficiency (SCID)-X1 disease. Science 288:669–672
8. Mi P, Cabral K, Kokuryo D et al (2013) Gd-DTPA-loaded polymer–metal complex micelles with high relaxivity for MR cancer imaging. Biomaterials 34:492–500
9. Tang Z, He C, Tian H et al (2010) Polymeric nanostructured materials for biomedical applications. Prog Polym Sci 60:86–128
10. Zhang W, Liao SS, Cui FZ (2003) Hierarchical self-assembly of nano-fibrils in mineralized collagen. Chem Mater 15:3221–3226
11. Takeuchi H, Yamamoto H, Kawashima Y (2001) Mucoadhesive nanoparticulate systems for peptide drug delivery. Adv Drug Deliv Rev 47:39–54
12. Liu T, Tang J, Jiang L (2004) The enhancement effect of gold nanoparticles as a surface modifier on DNA sensor sensitivity. Biochem Biophys Res Commun 313:3–7
13. Nakamura H, Karube I (2003) Current research activity in biosensors. Anal Bioanal Chem 377:446–468
14. Chen F, Zhang S, Bu W et al (2012) A uniform Sub-50 nm-sized magnetic/upconversion fluorescent bimodal imaging agent capable of generating singlet oxygen by using a 980 nm laser. Chem A Eur J 18:7082–7090
15. Guo H, Qian H, Idris NM et al (2010) Singlet oxygen-induced apoptosis of cancer cells using upconversion fluorescent nanoparticles as a carrier of photosensitizer. Nanomed 6:486–495
16. Pan Y, Zhang L, Zeng L et al (2015) Gd-based upconversion nanocarriers with yolk–shell structure for dual-modal imaging and enhanced chemotherapy to overcome multidrug resistance in breast cancer. Nanoscale 8:878–888
17. Cheng L, Yang K, Li Y et al (2011) Facile preparation of multifunctional upconversion nanoprobes for multimodal imaging and dual-targeted photothermal therapy. Ange Chem Int Ed 50:7385–7390

Chapter 14
Nanostructured Materials for Photonic Applications

Abstract The nanostructured materials play a significant role in the technological progress of photonic and optical applications. More photonic applications are based on functional nanostructures. The investigation on nonlinear properties are great importance in the photonic device fabrication. This chapter elaborates the functional nanomaterials and their properties for photonic applications.

Much of the existing attention in photonics arises due to the applicability of photonics to current and future information and image processing technologies. The realization of this advanced technology rests on the development of multifunctional materials which simultaneously satisfy many functional requirements. Synthesis, processing, and characterization of nanostructured materials may lead to numerous electronic and photonic applications. Nanostructured materials include nanowires, nanotubes, quantum dots, nanocrystalline and nanoparticle thin films, self-assembled and deposited organic nanofilms, etc. The electronic and photonic applications of nanostructured materials include electronic storage and computing, solar cells, photoluminescence, waveguide, chemical and biological sensing, and much more [1] (Fig. 14.1).

The excitement in the field of quantum computing was triggered in 1994 by Peter Shor who showed how a quantum algorithm could exponentially speed up a classical computation. Such algorithms are implemented in a device that makes direct use of quantum mechanical phenomena such as entanglement and superposition. Since the physical laws that govern the behavior of a system at the atomic scale are inherently quantum mechanical in nature, nanotechnology has emerged as the most appropriate tool to realize quantum computers [2].

Nanostructures based on organic optical materials, serving as fundamental elements in modern photonic devices, have attracted a great deal of research interest, due to the unique advantages of organic compounds, including high reaction activity, good processability, and high photoluminescence (PL) efficiency. One dimensional (1D) nanostructures have been demonstrated to be effective building blocks for miniaturized devices, because of the two-dimensional photon confinement. In addition, crystalline 1D organic nanostructure with the highly ordered stacking of conjugated molecules offer better stabilities and charge transport properties, which

T. D. Thangadurai et al., *Nanostructured Materials*, Engineering Materials,
https://doi.org/10.1007/978-3-030-26145-0_14

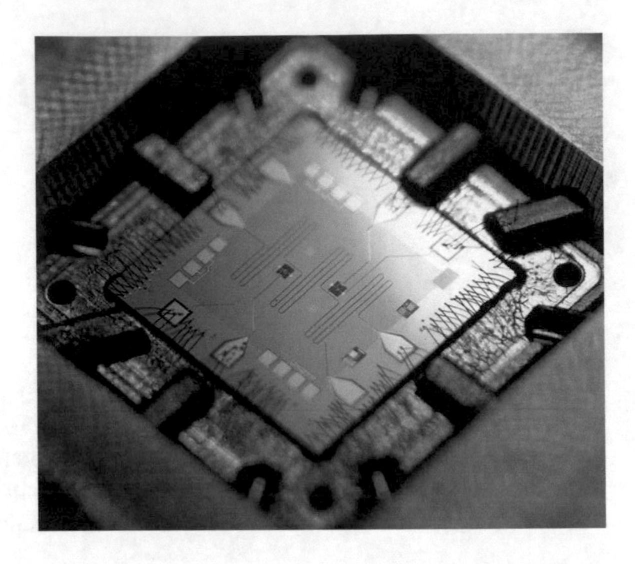

Fig. 14.1 Quantum computing chip: the two black squares are the quantum bits or cubits, the processing centre; the meandering line at the Centre is the quantum bus; and the lateral meandering lines are the quantum memory. Reprinted with permission from © Erik Lucero

result in better photonic performances of the final devices. Some unique applications, like tunable color displays, field-effect transistors, chemical sensors, optical waveguides, and lasers [3].

14.1 Optical Waveguides Based on Small Organic Molecules

One-dimensional optical waveguides is a key building block to generate and propagate light, have been attracting considerable attention. They can help in the manipulation of light guiding, localization, and enhancement within submicrometer volumes, and play an essential role in developing micro/nano scale photonic devices. In the past couple of years, inorganic semiconductor 1D nanomaterials have been successfully adopted as optical waveguides. In comparison with those of inorganic semiconductors, micro/nanostructures consisting of organic luminescent molecules with low molecular weights offer some unique advantages, such as diversity of optoelectronic properties and easy synthesis either by liquid phase or vapor phase methods. Due to the weak intermolecular interactions and excellent luminescent properties, organic micro/nano structures are expected to work as optical waveguides with distinguishing features and can be readily prepared with desired sizes ranging from hundreds of nanometers to several microns [4].

14.2 Optically Pumped Organic Lasers

Optical waveguides can lead to some properties of laser light, because waveguides have common laser geometries. 1D waveguiding nanostructures with optically flat end facets can function as miniature optical cavities, which apply feedback. As the light passes backwards and forwards, it is amplified by the stimulated emission, and if the amplification exceeds the losses of the resonator, lasing begins. Nanolasers with strong polarized output are able to offer great opportunities in future applications such as nanooptical routing, emission, detection, data storage, sensing and near-field optics. Inspired by the existing optically and electrically pumped inorganic semiconductor and polymer nanowire lasers, researchers show increasing interest in organic-based lasers, for their unique features, such as a high degree of spectral tunability, large stimulated emission cross-sections, and the potential for simple high-throughput fabrication. Moreover, the ease of fabrication has made it possible to produce 1D organic laser with a wide variety of optical microcavities. Therefore, the fabrication of 1D organic nano laser is of great scientific interest and technological significance [5].

14.3 3D Photonic Crystals

The three-dimensionally periodic structure has a complete band gap or frequency range in which light cannot propagate for any direction or polarization. Only very particular structures have this property. In general, the crystal must be made up of materials with relatively large difference in refractive index, such as silicon and air, to create strong enough scattering for a complete gap. In addition, the particular geometry must be chosen with care. The face-centered cubic (fcc) lattice, for example, is particularly favorable to the creation of band gaps. Due to its nearly spherical Brillouin zone, the partial band gaps at the corners of the 3D Brillouin zone tend to overlap. Two examples of 3D photonic crystals are shown in Fig. 14.2. The woodpile

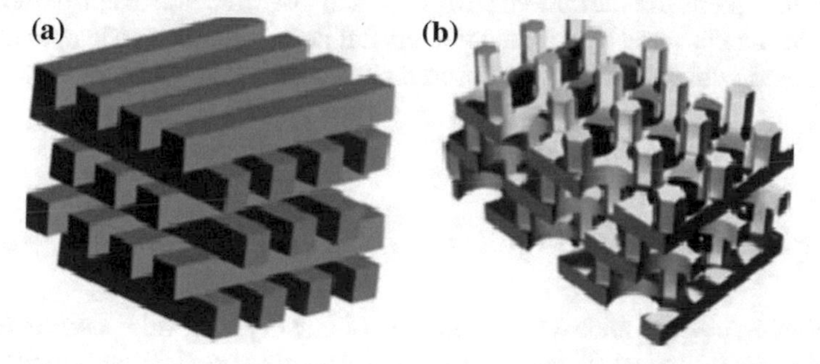

Fig. 14.2 Examples of 3D photonic crystals: **a** woodpile structure, **b** stacked rod and hole layer structure. Both belong to the fcc class of lattices

structure, shown in Fig. 14.2a, is made up of stacked layers of parallel rods with square cross-sections. Adjacent layers have perpendicular orientations.

The structure has a large photonic band gap of 17% of the mid-gap frequency for a silicon structure in air. The structure shown in Fig. 14.2b is made up of alternating layers of rods and holes. Each layer forms a triangular array. It also has a large photonic band gap of close to 20% for silicon in air. Because each of the layers resembles a 2D photonic crystal, the structure facilitates the design of waveguides and microcavities based on previously existing 2D designs. Because 3Dcrystals allow complete confinement of light in three dimensions, they may allow the design of complex, integrated optical circuits with unprecedented control over light flow [6].

14.4 Photonic LEDs

The interesting application of photonic crystals is light-emitting diodes (LEDs). Conventional semiconductor LEDs suffer from poor light extraction efficiencies due to total internal reflection at the semiconductor–air interface. For a semiconductor refractive index of $n = 3.4$ (corresponding to a half-cone angle of $\sin^{-1}(1/n) = 17°$), the efficiency is $1/4n^2 \sim 2\%$. Placing the active layer of the LED in a 2D photonic crystal slab can enhance the extraction efficiency. The photonic crystal should ideally be designed so that the LED emission spectrum falls in a frequency range for which all modes of the photonic crystal slab radiate to air. For a PC slab of sufficiently large area, close to 100% efficiency is expected theoretically. Experiments have observed a six-fold enhancement of photoluminescence in a InGaAs/InP double heterostructure. However, the penetration of air holes into the active layer of a LED causes additional surface recombination, lowering the internal quantum efficiency. An alternative scheme places a 2D photonic crystal grating above the active layer of a LED. If the photonic crystal layer is sufficiently shallow, the enhancement of the extraction efficiency can be described by the grating diffraction effect. Photonic crystal structures have also been used in vertical cavity surface emitting lasers (VCSELs). A photonic crystal was incorporated into the top distributed Bragg reflector (DBR), and an oxide aperture in the bottom DBR was used to restrict lateral current spreading beyond the area covered by the photonic crystal defect. Single-mode operation was demonstrated with sub-milliamp threshold current and a milliwatt of output power [7].

14.5 Photonic Crystal Filters

By combining cavities and waveguides in photonic crystals, it is possible to create different types of optical filters. Such devices may find application in optical communications, particularly in wavelength division multiplexing (WDM). In WDM

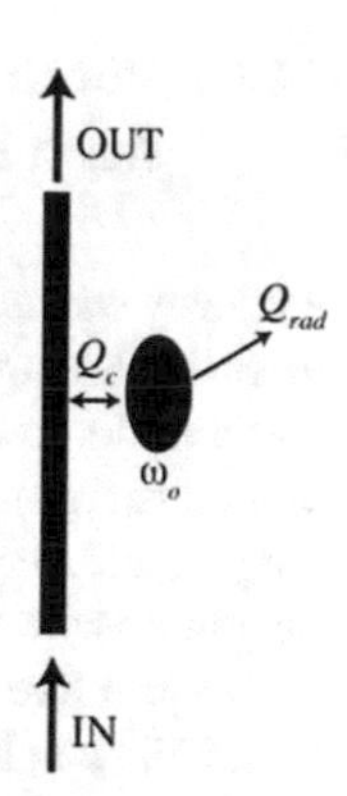

Fig. 14.3 Schematic diagram of band-rejection filter

systems, signals are encoded on multiple channels, each of which occupies a separate frequency bandwidth. Optical filters that can separate out and redirect particular channels from the optical data stream are useful for optical processing of the signal. In comparison to alternate technologies such as micro-ring resonators, photonic crystal devices are extremely compact in size, allowing denser on-chip integration. One basic filter design is shown schematically in Fig. 14.3. It consists of a waveguide side-coupled to a microcavity resonator. For simplicity, we represent the waveguide with a thick solid line and the microcavity resonator by a solid ellipse [8].

LC displays (LCDs) based on conventional nematic LCs now dominate the global market of advanced information displays with an annual worth of more than $100 billion, and have drastically revolutionized the way that information is presented. With LCDs ubiquitous in our daily life, research and applications of LCs are rapidly venturing into the forefront of the bottom-up nanofabrication of advanced photonic materials and devices, and promising new techniques have also been developed to controllably fabricate the liquid-crystalline nanostructures with tailored configurations. Different from the conventional photonic nanostructures fabricated by expensive, time-consuming, environmentally hazardous and scale-prohibitive manufacturing procedures of semiconductor based inorganic nanostructures, the unique self-organization properties inherent in the liquid-crystalline materials make it a quite advantageous approach toward the green, efficient, and cost-effective production of large-scale 3D periodic nanostructures. Thanks to the high flexibility of the self-assembly processes, unique building blocks and novel tuning mechanisms could be developed depending on the specific applications.

Importantly, being soft makes the liquid-crystalline materials responsive to various stimuli such as temperature, light, mechanical force, electric and magnetic field, and chemical and electrochemical reaction, resulting in tunable photonic band gaps in the 3D nanostructure. Compared with traditional 3D photonic crystals, the unrivalled attributes, such as multi stimuli-responsiveness, easy tunability, and real time configurability, 3D liquid-crystalline photonic superstructures would undoubtedly provide tremendous opportunities in the widespread applications of all-optical integrated circuits and next generation communication systems [9].

14.6 Photonic Crystals: Bright Structural Colour from Functional Morphology

Butterfly wings, mother-of-pearls and opals are examples of natural photonic structures displaying structural colour. This arises from interference between light waves and a periodic or random sub-wavelength medium and determines a range of forbidden energies in transmission, known as photonic bandgap. When the structures are periodic, these materials are called photonic crystals. Mimicking nature photonic structures with periodicity in one, two and three dimensions were realized. Several techniques like spin coating, electrochemical etching, and physical vapour deposition, PVD can be employed to fabricate one-dimensional photonic crystals. Refractive index contrast is obtained either varying the chemical composition of the layers or by controlling effective refractive index through the control of layers porosity. The realization of porous architecture represents a step forward in the exploitation of photonic crystals as it expands their potential function. Porous photonic crystals possess two great characteristics: (i) controlled mass transport of fluid analytes through interconnected porosity, and (ii) optical band structure engineering that enables the realization of tunable optical devices with potential applications in the fields of sensing, filtering, ICT (e.g. electro-optical switches), energy (sensitized solar cells), and photo-catalysis. The fabrication techniques of porous one-dimensional photonic crystals (p1DPCs) still lag behind the theoretical knowledge, and suffer for several limitations such as repeatability, scalability, and integrability [10].

The most common top-down fabrication technique for p1DPC is anodic current modulation during electrochemical etching of conductive wafers (e.g. doped-Si, Al). Pores size can be periodically changed from nanometers to micrometers obtaining a multilayer structure. This wet technique is limited to conductive materials and requires complex and multi-step fabrication achieving low refractive index contrast. To overcome this criticism a bottom-up assembly methods have been proposed to demonstrate what previously demonstrated by the theory. A number of different materials in nanoparticles colloidal dispersions were employed as p1DPC building blocks. Strong photonic bandgap efficiencies were obtained by using NP's material with intrinsically high refractive index and tuning particles size distribution with thickness of several micrometers (using two different materials) or of few millimeter (using structures based on opals). Spin coating processes, a cheap and flexible technique, was the preferred technique for the fabrication of these of devices. p1DPC was also fabricated from the gas phase via glancing angle deposition, a PVD technique capable of engineering different films architectures through shadowing of vapor atoms at highly oblique impact angles. Control over the porosity and thickness of individual layers is granted with the possibility to use different materials. The pulsed laser deposition is proposed as a novel fabrication technique for gas-phase self-assembled photonic hierarchical nanostructures. As the beetle cuticle, self-assembly of different hierarchical porous materials is used to achieve refractive index variations and in turn the photonic bandgap. Hierarchical one dimensional photonic crystal (h1DPC) with 75% reflection efficiency in a 0.5 μm thick device and approaching complete reflection in a 1.5 μm thick device was fabricated using a single material (TiO_2). Structural density of hierarchical nanostructures can be controlled allowing fabrication of layers having different optical and morphological properties. When stacked onto each other they achieve peculiar optical functionalities. The hierarchical photonic nanostructures respond to fluid infiltration with photonic bandgap red shifting. PLD allow the realization of patterned photonic devices, on the scale of few μm, with interesting perspectives in the fields of solar cells, biochemical sensing, photo electrochemical water splitting, displays and photo-catalysis. The characteristic low thermal load of PLD process allows using even plastic or biological substrates. PLD decouples the chemistry from morphological parameters of deposited nanostructures that are controlled by the background gas pressure. Refractive index contrast is obtained stacking layers of hierarchical nanostructures grown at pressures [11].

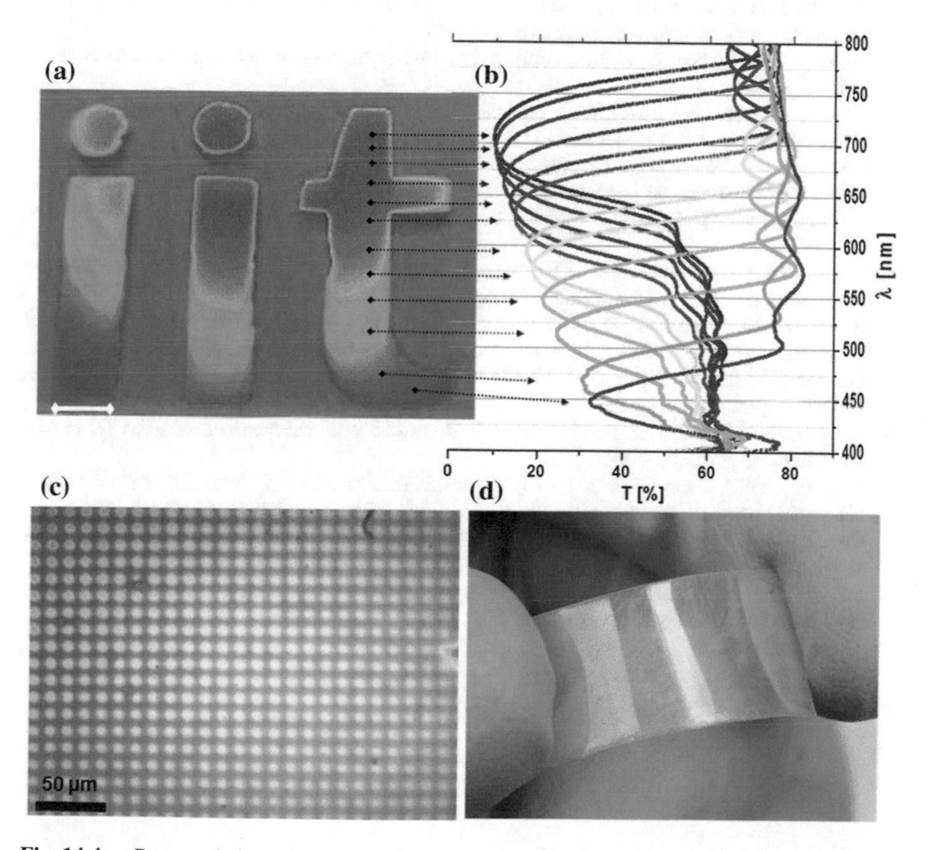

Fig. 14.4 **a** Patterned photonic crystal, fabricated with thickness gradient of the layers, by using a mask. Scale bar: 1 cm. **b** Transmission spectra of the photonic crystal along one of the letter, from top to bottom; a shift of the photonic band gap spectral position of few hundred of nanometers is shown. **c** Pixel structure of hierarchical photonic crystal. d) h1DPC deposited on flexible plastic substrate. Reprinted with permission from © ACS Nano, 8 (2014)

With PLD is possible to realize patterned porous photonic crystals. In Fig. 14.4a an example is reported where a h1DPC, placed off-axis with respect to the deposition source, was physically masked in such a way to obtain a letter pattern with a thickness gradient giving rise to a PBG red shift (from bottom to top) (Fig. 14.4b). Using the same approach a pixel structure of h1DPC (pixel size of about 10×10 μm) was realized (Fig. 14.4c). The pixel dimension is matching the requested pixel size in display technology and it could be appealing for achieving special sensing resolution. The large oriented and interconnected porosity of h1DPC can be used for measuring refractive index of fluids, with possible applications in optoelectronic sensors [12].

References

1. Xia YN, Yang PD, Sun YG et al (2003) One-dimensional nanostructures: synthesis, characterization, and applications. Adv Mater 15:353–389
2. Samkharadze N, Zheng G, Kalhor N et al (2018) Strong spin-photon coupling in silicon. Science 359:1123–1127
3. Zhao YS, Wu JS, Huang JX (2009) Vertical organic nanowire arrays: controlled synthesis and chemical sensors. J Am Chem Soc 131:3158–3159
4. Guo X, Qiu M, Bao JM et al (2009) Direct coupling of plasmonic and photonic nanowires for hybrid nanophotonic components and circuits. Nano Lett 9:4515–4519
5. Duan XF, Huang Y, Agarwal R et al (2003) Single-nanowire electrically driven lasers. Nature 421:241–245
6. Johnson SG, Joannopoulos JD (2000) Three-dimensionally periodic dielectric layered structure with omnidirectional photonic band gap. Appl Phys Lett 77:3490–3492
7. Boroditsky M, Krauss TF, Coccioli R et al (1999) Light extraction from optically pumped light-emitting diode by thin-slab photonic crystals. Appl Phys Lett 75:1036–1038
8. Sharkawy A, Shi SY, Prather DW (2001) Multichannel wavelength division multiplexing with photonic crystals. Appl Opt 40:2247–2252
9. Bisoyi HK, Li Q (2014) Light-directing chiral liquid crystal nanostructures: from 1D to 3D. Acc Chem Res 47:3184–3195
10. Yang Q, Zhu S, Peng W et al (2013) Bioinspired fabrication of hierarchically structured, pH-tunable photonic crystals with unique transition. ACS Nano 7:4911–4918
11. Guo DL, Fan LX, Wang FH et al (2008) Porous anodic aluminum oxide bragg stacks as chemical sensors. J Phys Chem C 112:17952–17956
12. Kobler J, Lotsch BV, Ozin GA et al (2009) Vapor-sensitive Bragg mirrors and optical isotherms from mesoporous nanoparticle suspensions. ACS Nano 3:1669–1676

Chapter 15
Nanostructured Materials for Environmental Remediation

Abstract Currently, more research as concerned on environmental remediation for improving our daily lives and the environment. The nanoscale materials designed for environmental application has more concern towards increased surface area, surface modification, and tunability of size. This review mainly describes recent progress in the design, fabrication, and modification of nanostructured semiconductor materials for environmental applications.

Nanostructures have several advantages over their bulk counterparts that could potentially be exploited for environmental remediation. For example, the higher surface area-to-volume ratio of nanostructures could lead to an enhanced reactivity with environmental contaminants that degrade through inner-sphere adsorption mechanisms. Nanostructured materials feature with controllable size, shape, composition, structure, and surface, which are difficult or impossible to achieve in their bulk materials. Such unique features of nanostructured materials can be further tailored and engineered to specifically tackle a particular energy and/or environmental challenge [1]. Also the nanostructured materials provide more flexible space for ease reconstruction, as their nanosize expands the limits and results in confinement effect, enhanced mechanical stability, and large surface area, and make them suitable for photocatalytic activities. The advancement in synthesis techniques provides the freedom to alter its physical properties as per the demand [2].

The development of nanostructured materials has opened new paths in different scientific fields and is providing new opportunities in environmental science. Advanced oxidation processes have benefited from the development of nanostructured semiconductors giving rise to heterogeneous photocatalysis, which can achieve total mineralization of dissolved organic contaminants in both aqueous media, such as in air, in less time and without the addition of sacrificial oxidants (H_2O_2, O_3, etc.) [3]. Nanoscale iron particles may offer real advantages over granular iron because they provide a greater surface area and therefore higher reactivity. Metallic nanoparticles can be applied through direct injection of the particles to contaminated sediments and aquifers. Surface modification of metallic nanoparticles gives the added advantage of protecting the surface from oxidation, thus improving the long-term activity.

T. D. Thangadurai et al., *Nanostructured Materials*, Engineering Materials,
https://doi.org/10.1007/978-3-030-26145-0_15

The nanostructured materials designed for metal ions trapping and absorbing has received great research interest and many typical nanostructure systems have been demonstrated since the end of last century. Basically, the advantages nano could provide include high surface to bulk ratio associated with high porosity, large metal ion absorbing capacity, and good regeneration ability. The selection and design for absorbent materials are required to meet several criteria as follows. First, the absorbent itself should be non-toxic in nature. Second, the absorbent should have selectivity and sensitivity as high as possible toward specific type of ion in addition to the large adsorption capability. Third, the adsorbed ions must be easy to collect off the surface of absorbent materials so that the absorbents could be used for repeated times. Many metal oxides are low cost and non-toxic materials and the hierarchical nature makes them promising for ion removal in the waste water. Typical examples include magnesium oxide (MgO), aluminum oxide (Al_2O_3), iron oxide (Fe_2O_3), titanium oxide (TiO_2) and so on. Much research progress has been made in the synthesis of these metal oxides with controlled nanoscale morphology. Hierarchical iron oxide nanostructures have been extensively studied among all the metal oxide nanostructures for the heavy metal ion removal. In addition to the large adsorption capacity, high surface area and low cost, an important advantage over other metal oxides is the magnetic property which makes it easily recoverable from water after toxic ion removal. Magnetite (Fe_3O_4) and magnetite (γ-Fe_2O_3), which are ferromagnetic and ferromagnetic in nature, have been proved to act as efficient ion absorbents [4].

15.1 Gas Treatment: Nano-Array Based Catalytic Converters

The air pollution has become a world-wide crisis resulting from burning of fossil fuels. A major portion of the air pollutants comes from the automobile exhaust and power plant. The waste gas usually constitutes nitrogen oxide and sulfur dioxide, which either triggers photochemical pollution or contributes to acidic rain as well as hydrocarbon and carbon dioxide, which lead to green-house effect. Specific catalyst is needed to deal with each pollution gas component. The catalytic techniques are in the progress for the preparation of nano catalyst toward different gas treatment [5].

15.2 Remediation of Organohalides by Dye Sensitized TiO_2

Dye sensitization has been frequently used to extend the photoresponse of TiO_2 into the visible region. Photoexcitation of a sensitizer molecule on the surface of TiO_2 can result in electron injection into TiO_2. This mechanism is illustrated in Fig. 15.1 and produces a TiO_2 (e^-)–S* charge separated state that can in principle reduce and/or oxidize environmental pollutants [6].

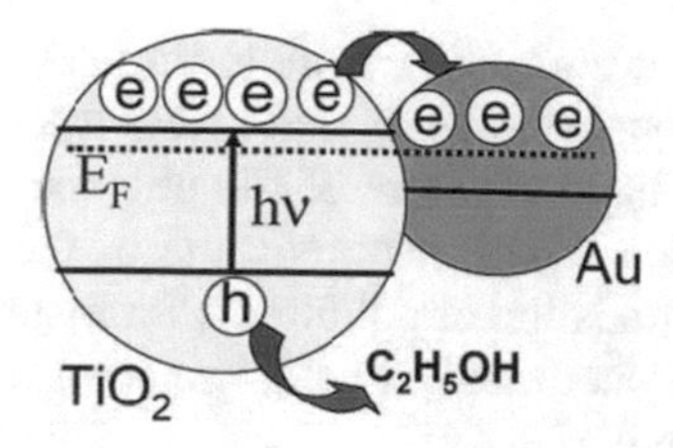

Fig. 15.1 Charge distribution in semiconductor-metal composite system leads to fermi-level equilibration

15.3 Water Split Application of the Nanostructure Photocatalyst

Titanium oxide is a most studied photocatalyst as it has low toxicity, high chemical and biological inertness, and high photocatalytic reactivity. Its limitation is that it is naturally activated by UV radiations which form a very low percent of solar spectrum. However, doping with transitional metals such as Pt, V, Cr, or Fe have enabled the absorption spectrum towards visible. In a similar effort, hierarchal porous structure (HPT-500) TiO_2 was deposited with Pt and Zinc phthalocyanine (PCH-001) to form a composite. The results showed that Pt-HPT nanocomposite with 0.25 μmol/15 mg of nanocomposite (HPT-0.25) exhibited maximum H_2 evolution. The H_2 yield was found to be 2260 μmol at $\lambda \geq 450$ nm and the turn over number was found to be 18,080. Highest apparent quantum yield (AQY) of 11.97% was achieved at 690 nm for HPT-0.25. Exfoliation is a technique that can either modify or reduce the size of the photocatalyst. One major advantage of using exfoliation is that it does not alter the crystal structure. Due to reduction in size, the migration distance of charge carriers is reduced and the surface area of the catalyst is increased new ultrasonic method for rapid and high yield of (TBA/H) $Pb_2Nb_3O_{10}$ by exfoliation. The obtained (TBA/H) $Pb_2Nb_3O_{10}$ was loaded with Pt and used as visible light photocatalyst for production of hydrogen ($\lambda > 415$ nm), thereby making it as a first and foremost visible light-driven $AB_2Nb_3O_{10}$. The perovskite yielded 50 mmol/h and 17 mmol/h of H_2 under full-arc and visible light irradiation, respectively. The study concluded that, due to smaller migration distance of charge carriers, efficiency of electrons for hydrogen production improved [7].

15.4 Nanostructure Photocatalyst for Water and Wastewater Treatment

Nanostructured photocatalytic materials are broadly used in wastewater purification and for degrading the recalcitrant organics. The most common materials used are TiO_2, ZnO, WO_3, g-C_3N_4, etc. Photocatalyst have proven to be helpful in degrading several categories of organic contaminants including POPs and endocrine disrupting compounds (EDCs). Various parameters that affect the photocatalysis include pH, availability of solar light, band structure of the photocatalyst, concentration of the photocatalyst, concentration of pollutant, etc. pH of the reaction mixture exhibits profound effect on the degradation of pollutants, for instance, in an experiment performed was found that the photocatalyst heterojunction (p-BiOI/p-NiO) was able to degrade acid orange 7 more effectively at strong acidic condition as compared to neutral. The acidic condition modified the surface charge of the photocatalyst and thus enhanced the degradation process. Other important parameters that can affect the degradation process are photocatalyst dosage and pollutant concentration [8].

Nanostructured membranes (NSM) are used to reject suspended particles, bacteria, macromolecules, viruses, colloids, organic compounds, and multivalent ions. They are mainly utilized for drinking water production for water softening or disinfection, and the treatment of landfill leachate and industrial wastewater. The advantages of NSM compared with other treatment methods depend upon the substances present in the feed water and the water that is subject to cleaning. One advantage is that different pollutants can be removed in the same filtration step. Depending on the membrane this removal can be selective towards certain substances or compounds can be only partially removed; for example calcium carbonate in the case of water softening. It is therefore crucial that the membrane is exactly tailored to the needs of the treatment plant in order to achieve the maximum removal of the target compounds. Additional advantages are non-resistant organisms are left, the use of chemicals is minimal, and the cleaning capacity is not detracted by turbidity, or economic factors such as costs or energy use.

The development of nanostructured solar energy driven semiconductor photocatalyst have received immense attention in the areas of environmental remediation and energy storage. The numerous organic pollutants are discharged into the environment by an increasing revolution in industrial and agricultural sectors. The presence of organic pollutants in the wastewater can change the oxygen levels that can drastically lead to severe problems in the ecosystem. The photocatalytic degradation is considered to be an effective ways for the removal of organic pollutants from wastewaters. Therefore, it is necessary to develop highly effective, visible light active catalytic materials to solve most of the critical problems related to energy and environment [9].

Detection of toxic gases in the environment is an important and critical aspect to industry, agriculture and human health. It is essential to develop a reliable and selective gas sensor for determining toxic gases. Among various gas sensors, semiconductor-based gas sensors have attracted much attention because of the advantages of convenience, inexpensive, and rapid detection. More recently, MoS_2 has been widely investigated by a number of researchers in gas sensing applications. The gas molecules can be easily adsorbed on the surface of MoS_2. On the other hand, the weak interaction allows gas atoms or electrons to infiltrate and transport freely in MoS_2. Therefore, MoS_2-based nanostructured materials has turned into potential candidates for the fabrication of high-performance gas sensors, such as NH_3, H_2, NO, and many kinds of chemical vapors [10].

15.5 Sensing the Chemical Environment with Semiconductor Nanostructures

The sensors for environmental applications have been continuously developing, which includes, SnO_2 based semiconductor systems that have been used as conductometric gas sensors and a TiO_2 electrode for determining the chemical oxygen demand (COD) of water. Prompted by the discovery of photoluminescence from porous Si and subsequent observation that many organic and inorganic molecules quench this emission, has led to their exploitation as chemical sensors. The photoluminescence in porous Si arises from charge carrier recombination in the quantum confined nanosized environments. The feasibility of using porous silicon nanocrystals as sensors has been demonstrated for detecting nerve gas agents and nitro organics. Using detection based on electrical conductivity, photoluminescence, or interferometry, researchers have achieved sensitivity in the range of ppb–ppm levels [11].

15.6 Pollution Control Using Nanostructures

15.6.1 Air Pollution

Air pollution can be remediated using nanotechnology; for example, the use of nanocatalysts with increased surface area for gaseous reactions. Catalysts work by speeding up chemical reactions that transform harmful vapors from cars and industrial plants into harmless gases. Catalysts currently in use include a nanofiber catalyst made of manganese oxide that removes volatile organic compounds from industrial smokestacks [12]. Another approaches are, using nanostructured membranes that have small pores to separate methane or carbon dioxide from exhaust. The carbon nanotubes (CNT) for trapping greenhouse gas emissions caused by coal mining and power generation. CNT can trap gases up to a hundred times faster than other methods, allowing integration into large-scale industrial plants and power stations. This new technology both processes and separates large volumes of gas effectively, unlike conventional membranes that can only do one or the other effectively.

15.6.2 Water Pollution

The harmful pollutants in water can be converted into harmless chemicals through chemical reactions. Trichloroethene, a dangerous pollutant commonly found in industrial wastewater, can be catalyzed and treated by nanoparticles. Researchers have shown that these materials should be highly suitable as hydrodehalogenation and reduction catalysts for the remediation of various organic and inorganic groundwater contaminants [13]. Nanotechnology used for water cleansing process because inserting nanoparticles into underground water sources is cheaper and more efficient than pumping water for treatment. The deionization method of using nano-sized fibers as an electrode is cheaper, and energy efficient. Traditional water filtering systems use semi-permeable membranes for electrodialysis or reverse osmosis. Decreasing the pore size of the membrane to the nanometer range would increase the selectivity of the molecules allowed to pass through. Membranes that can even filter out viruses are now available. Also widely used in separation, purification, and decontamination processes are ion exchange resins, which are organic polymer substrate with nano-sized pores on the surface where ions are trapped and exchanged for other ions.

Ion exchange resins are mostly used for water softening and water purification. In water, poisonous elements like heavy metals are replaced by sodium or potassium. However, ion exchange resins are easily damaged or contaminated by iron, organic matter, bacteria, and chlorine. Recent developments of nano-wires made of potassium manganese oxide can clean up oil and other organic pollutants while making oil recovery possible. These nanowires form a mesh that absorbs up to twenty times its weight in hydrophobic liquids while rejecting water with its water repelling coating. Since the potassium manganese oxide is very stable even at high temperatures, the oil can be boiled off the nanowires and both the oil and the nanowires can then be reused [14].

Designing new catalysts that can efficiently utilize multiple energy sources can contribute to solving the current challenges of environmental remediation and increasing energy demands. The fabricated single-crystalline $BiFeO_3$ (BFO) nanosheets and nanowires that can successfully harness visible light and mechanical vibrations and utilize them for degradation of organic pollutants. Under visible light both BFO nanostructures displayed a relatively slow reaction rate. However, under piezocatalysis both nanosheets and nanowires exhibited higher reaction rates in comparison with photocatalytic degradation. When both solar light and mechanical vibrations were used simultaneously, the reaction rates were elevated even further, with the BFO nanowires degrading 97% of Rhodamine B dye within 1 h (k-value 0.058 min^{-1}). The enhanced degradation under mechanical vibrations can be attributed to the promotion of charge separation caused by the internal piezoelectric field of BFO. BFO nanowires also exhibited good reusability and versatility toward degrading four different organic pollutants [15] (Fig. 15.2).

15.7 Field-Effect Transistor (FET) Sensors

FET sensors have been used for detection of heavy metals by utilizing the interaction between the analyte and semiconductor resistor. FET sensors are capable of real-time label-free detection of heavy metals. One-dimensional and two-dimensional semiconducting nanomaterials are especially attractive to the FET sensors because they have very high surface-to-volume ratio, leading to high sensitivity. A Si nanowire was used to construct a FET sensor for detection of heavy metals. This FET sensor

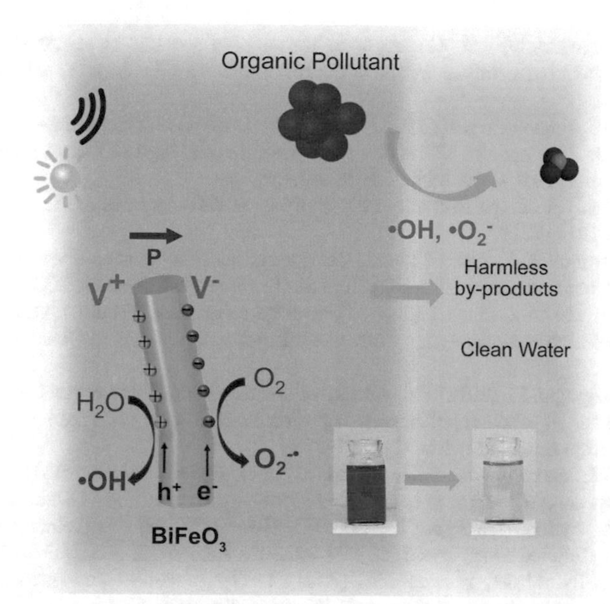

Fig. 15.2 Catalytic degradation schemes of BFO NWs under light and ultrasound. Reprinted with permission from © Science, 29, 236–246 (2018)

exhibited high sensitivity with a LOD of 10^{-7} M for Hg^{2+} and 10^{-4} M for Cd^{2+}. Another example was a selective and sensitive FET sensor constructed with single walled carbon nanotubes (SWCNTs), which was based on the conductance change due to the selective redox reaction between SWCNTs and Hg^{2+}. Reduction from Hg^{2+} to Hg^0 by the SWCNTs is thermodynamically favorable, while reduction of other metal ions with SWCNTs is unfavorable due to their negative potentials [16].

Also, Nanostructured porous Si (PSi) has been recognized as a versatile platform for numerous sensing and biosensing applications, mainly for its tunable optical properties and large surface area. PSi-based interferometers are highly sensitive to the presence of chemical or biological molecules within the pores, due to the change in the average refractive index of the nanostructure. Ultimately, the porous scaffold offers an unbiased label-free optical detection of a wide variety of biomolecular interactions, e.g., enzyme-substrate, antibody-antigen4 and DNA fragments, which are facilitated over a small working area [17].

References

1. Xia H, Hu MZ, ShirleyMeng Y et al (2014) Nanostructured materials for clean energy and environmental challenges. J Nanomater 675859:2
2. Rani A, Reddy R, Sharma U et al (2018) A reviews on the progress of nanostructure materials for energy harnessing and environmental remediation. J Nanostruct Chem 8:255–291

3. Rodriguez-Gonzalez V, Ruiz-Gomez MA, Torres-Martinez LM et al (2009) Sol–gel silver hexatitanates as photocatalysts for the 4-chlorophenol decomposition. Catal Today 148:109–114
4. Chen D, Cao L, Hanley TL et al (2012) Facile synthesis of monodisperse mesoporous zirconium titanium oxide microspheres with varying compositions and high surface areas for heavy metal ion sequestration. Adv Funct Mater 22:1966–1971
5. Rodhe H (1990) A comparison of the contribution of various gases to the greenhouse effect. Science 248:1217–1219
6. Jacob M, Levanon H, Kamat PV (2003) Charge distribution between UV-irradiated TiO_2 and gold nanoparticles: determination of shift in the fermi level. Nano Lett 3:353–358
7. Maeda K, Eguchi M, Oshima T (2014) Perovskite oxide nanosheets with tunable band-edge potentials and high photocatalytic hydrogen-evolution activity. Angew Chem Int Ed 53:13164–13168
8. Yosefi L, Haghighi M (2018) Fabrication of nanostructured flowerlike p-BiOI/p-NiO heterostructure and its efficient photocatalytic performance in water treatment under visible-light irradiation. Appl Catal B Environ 220:367–378
9. Theerthagiri J, Senthil RA, Senthilkumar B et al (2017) Recent advances in MoS nanostructured materials for energy and environmental applications—a review. J Solid State Chem 252:43–71
10. Donarelli M, Prezioso S, Perrozzi F et al (2015) Response to NO_2 and other gases of resistive chemically exfoliated MoS_2-based gas sensors. Sens Actuators B 207:602–613
11. Germanenko IN, Li ST, El-Shall MS (2001) Decay dynamics and quenching of photoluminescence from silicon nanocrystals by aromatic nitro compounds. J Phys Chem B 105:59–66
12. Bhawana P, Fulekar MH (2012) Nanotechnology: remediation technologies to clean up the environmental pollutants. Res J Chem Sci 2:90–96
13. Nutt MO, Hughes JB, Wong MS (2005) Designing Pd-on-Au bimetallic nanoparticle catalysts for trichloroethene hydrodechlorination. Environ Sci Technol 39:1346–1353
14. Yuan J, Liu X, Akbulut O et al (2008) Superwetting nanowire membranes for selective absorption. Nat Nanotechnol 3:332–336
15. Mushtaq F, Chen X, Hoop M et al (2018) Piezoelectrically enhanced photocatalysis with $BiFeO_3$ nanostructures for efficient water remediation. Science 4:236–246
16. Li M, Gou H, Al-Ogaidi I (2013) Nanostructured sensors for detection of heavy metals: a review. ACS Sustain Chem Eng 1:713–723
17. Shtenberg G, Massad-Ivanir N, Segal E (2015) Detection of trace heavy metal ions in water by nanostructured porous Si biosensors. Analyst 140:4507–4514

Chapter 16
Miscellaneous Applications of Nanostructures

Abstract Other than energy, environment, electronic and biomedical applications, the nanostructured materials used in more advanced and sensitive applications like cosmetics, food processing, communication and security purpose. This chapter describes all the applications related to above mentioned sectors.

Nanostructured materials for regenerative medicine also include particles, which can be used as an important tool for both sensing and actuating tissue morphogenesis. For example, particles can be used as functional components for biosensors and cell tracking devices to characterize nascent tissue morphogenesis in vivo. Nanoparticles can also serve as critical biomaterials to stimulate in situ tissue morphogenesis by virtue of such strategies as receptor-ligand interactions, controlled release, and gene delivery. The use of nanoparticle-based gene delivery vectors study concludes that uptake of functionalized particles is primarily dependent upon nonspecific interactions with heparin sulfate in conjunction with integrin binding domains [1].

16.1 Nanostructures in Aerospace Application

The aerospace applications for nanotechnology include high strength, low weight composites, improved electronics and displays with low power consumption, variety of physical sensors, multifunctional materials with embedded sensors, large surface area materials and novel filters and membranes for air purification, nanomaterials in tires and brakes and numerous others. The opportunities for aerospace industry are through thermal barrier and wear resistant coatings, sensors that can perform at high temperature and other physical and chemical sensors, sensors that can perform safety inspection cost effectively, quickly, and efficiently than the present procedures, composites, wear resistant tires, improved avionics, satellite, communication and radar technologies [2].

T. D. Thangadurai et al., *Nanostructured Materials*, Engineering Materials,
https://doi.org/10.1007/978-3-030-26145-0_16

16.2 Nanostructures in RADAR Application

Nanostructured polymer composites have released up new views for multifunctional materials. In particular, carbon nanotubes (CNTs) present potential applications in order to improve mechanical and electrical performance in composites with aerospace application. The combination of epoxy resin with multiwalled carbon nanotubes results in a new functional material with enhanced electromagnetic properties. The processing of radar absorbing materials based on formulations containing different quantities of carbon nanotubes in an epoxy resin matrix. The adequate concentration of CNTs in the resin matrix was determined and the processed structures were characterized by scanning electron microscopy, rheology, thermal and reflectivity in the frequency range of 8.2–12.4 GHz analyses. The microwave attenuation was up to 99.7%, using only 0.5% (w/w) of CNT, showing that these materials present advantages in performance associated with low additive concentrations [3]. Radar Absorbing Materials (RAM) provides energy losses of electromagnetic radiation. In certain frequency bands, these materials attenuate the incident electromagnetic wave radiation and dissipate the energy absorbed in the form of heat through internal mechanisms, magnetic and/or dielectrics. These loss mechanisms can be physical, chemical or simultaneously both. Increased electromagnetic pollution due to the presence of microwaves and the use of stealth technology in defense systems and military platforms have been the major attractions for studies in this area, with investments in research that already cover the frequency range of 1–40 GHz. Some recent studies have described radar absorbing structures applied in the frequency range of 10–100 GHz [4].

16.3 Nanostructures in Stealth Application

Block copolymer micelle lithography (BCML) is a potent method for preparing periodically spaced metallic nanoparticles with sub-100 nm spacing on different substrates. Highly ordered arrays of nanoparticles are fabricated using BCML and used as etching masks in a reactive ion etching (RIE) process to obtain well-defined arrays of nanopillars perpendicular to the interface. These nanopillars gradually adjust the difference in the refractive indices of the substrate material and the surrounding medium (e.g., vacuum, air, or water). The combination of BCML and RIE is a reflective method of choice for producing surfaces for antireflection applications in the ultraviolet (UV) and visible range. Using this physical principle enables increased light transmittance, reduced reflectance, and low absorption to be obtained within a range of wavelengths covering several hundred nanometers. The wavelength of maximal transmittance and minimal reflectance depends on the effective refractive index and the optical thickness of the layer that is made up of the nanopillar array [5].

One strategy to tune the effective refractive index of this interface layer within a limited wavelength range is to adjust the spatial spacing and the width of the nanopillars. The nanostructured stealth surface, with minimal reflectance (<0.02%) and maximal transmittance (>99.8%) for a wavelength range, covering visible and near-infrared. Compared to multilayer thin film coatings for near-infrared applications our antireflective surfaces operate within a much broader wavelength range, are mechanical stable to resist human touch or contamination, show a 44% higher laser-induced damage threshold, and are suitable for bended interfaces such as microlenses as well [6].

16.4 Nanostructured Electrode

Mostly, the lithium–air batteries have been described to have a higher specific capacity than most other batteries systems, but the rate capability, cycle life, and power performance of lithium–air batteries are still not satisfied for practical applications. Some of the major obstacles are limiting oxygen solubility and diffusion, accumulation of reaction products, and the lack of effective 3-phase electrochemical interface, which are directly determined by electrode structure design. A porous electrode with optimum porosity and effective catalysis site distribution for maximization of materials utilization is desirable. In addition to optimizing the air electrode composition which may affect the cathode porosity, more efforts should be devoted towards designing a novel porous air electrode. The simulated and analyzed several air electrode designs including single pore system, dual pore system in two dimensions, and dual pore system with multiple time release catalysts.

Some important parameters including the porosity distribution, pore connectivity, the tortuosity of the pore system, and the catalyst spatial distribution were studied in detail [7]. The results indicated that the dual pore system offers advantages for improving oxygen transport into the inner regions of the air electrode. When coupled with multiple time-release catalysts, the system can substantially extend the duration at higher powers, and result in maximum utilization of air electrode materials. Considering the requirement of porous structure and available void volume for discharge products, numerous efforts have been devoted to designing a novel porous air electrode with maximum void volume. Researchers developed all-carbon–nanofiber porous electrode with highly efficient utilization of carbon material and void volume for lithium–air batteries which was found to yield high gravimetric energies of four times higher than lithium-ion batteries. A similar freestanding carbon nanotube/nanofiber mixed buckypaper was also developed and applied in lithium–air batteries [8] (Fig. 16.1).

The chemical-biochemical functionalization and characterization of devices based on nanostructured materials i.e. porous silicon (PSi), zinc oxide nanowires (ZnO NWs) and diatomite for abovementioned applications. A valid chemical functionalization strategy is crucial for the realization of devices capable of selective detection

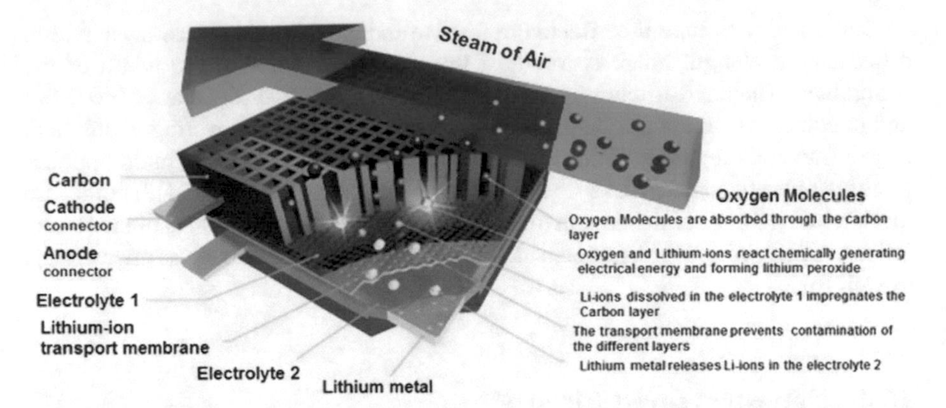

Fig. 16.1 The structure design of an air-breathing lithium–air battery developed by IBM. Reprinted with permission from © Nano Energy, 2, 443–467 (2013)

due to the combination of specificity of biological recognition probes and the sensitivity of sensor, providing unambiguous identification and accurate quantification. A common strategy based on the salinization process by aminosilane compound (e.g. APTES, APDMES) was carried out, resulting a valid preliminary chemical treatment to passivate and/or functionalize nanostructured surface in order to covalently bind biological molecules such as single or double stranded DNA, proteins, enzymes, antibodies, aptamers and so on, acting as bioprobes for biosensor realization. Nanostructured photonic materials are optimal transducers for optical sensing due to their capacity to convert molecular interactions in light signals without contamination or deterioration of the samples [9] (Fig. 16.2).

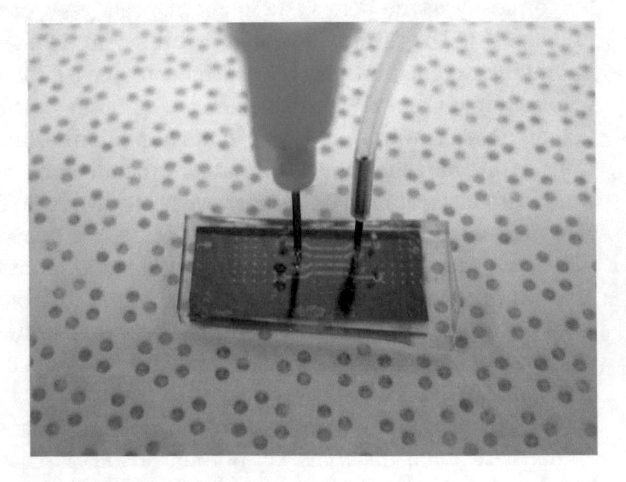

Fig. 16.2 Image of the porous silicon microarray integrated with the microfluidic circuit

16.5 Nanostructures in Antimicrobial Application

Nanostructured materials (NSMs) have progressively used as a ancillary for antibiotics and additives in various products to impart microbicidal effect. In particular, use of silver nanoparticles (AgNPs) has gathered enormous researcher's attention as potent bactericidal agent due to the inherent antimicrobial property of the silver metal. Moreover, other nanomaterials carbon nanotubes, fullerenes, graphene, chitosan, etc., have also been studied for their antimicrobial effects in order ensure their application in widespread domains. Antimicrobial effect of NSMs has been widely studied by several research groups against a wide range of microorganisms. NSMs can be regarded as the next generation antibiotics as they possess remarkable potential to overcome multidrug resistance problems in the pathogenic microbes. Depending on their ability to provide biostatic and biocidal action against microbial species, they can also be exploited in healthcare and personal care products, food safety, crop protection, water treatment, textile industries, etc. Although NSMs have shown spectacular antimicrobial effect against more than 500 microbial species, however, accurate mechanism behind their microbicidal activity is not hitherto well-understood [10] (Fig. 16.3).

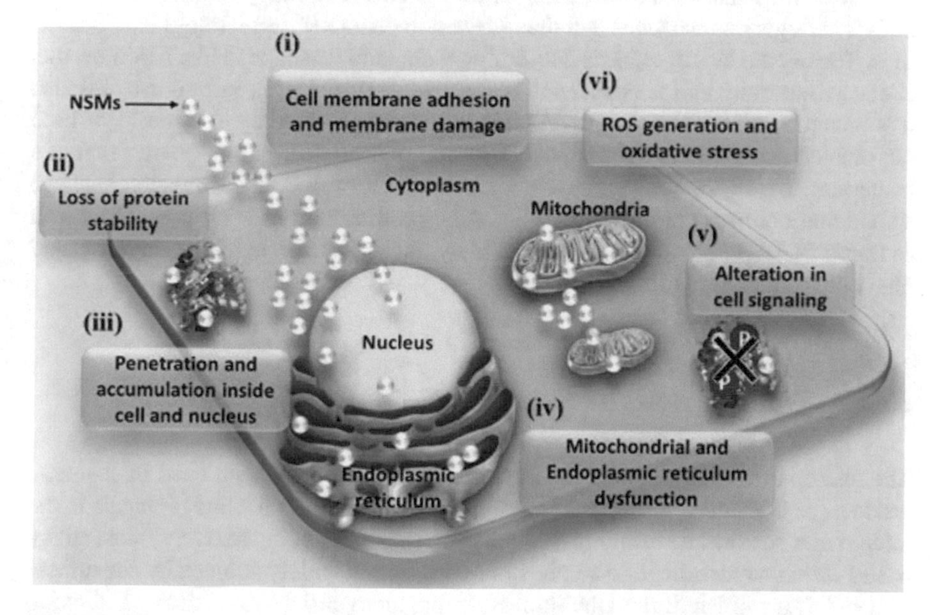

Fig. 16.3 The various modes of microbial toxicity caused by nanostructured materials

16.6 Nanostructures in Cosmetic Application

The cosmetics industry uses nanodispersions encapsulation or carrier systems, so that arbitrators enter into deeper skin layers where they activate skin metabolism with the aim of improving the skin's appearance. The functions and benefits of these encapsulation and carrier systems are, the controlled release and optimization of the availability of cosmetic agents in certain skin layers, the protection of sensitive agents, longer shelf life and hence greater product effectiveness, and a reduction in the amount of agents and additives used in products. They include liposomes, nanoemulsions, microemulsions and lipid-nanoparticles which are soluble and biodegradable [11].

16.6.1 Chitin Nanofibril

A chitin nanofibril is a nanocrystal of a natural polysaccharide obtained from the crustacean exoskeleton after elimination of the carbonate and protein portions. Having a backbone like hyaluronic acid, chitin nanofibril is easily metabolized by the body's endogenous enzymes and thus is used in cosmetic dermatology and biotextiles. The crystal average size is $240 \times 7 \times 5$ nm, and its shape is like a thin needle. It occurs naturally and is considered a safe raw material, it is safe to use and also it is both bio and eco compatible. As it is easily metabolized by enzymes, it is both bio- and eco-compatible. As the nanofibril has an average size one-quarter that of a bacterium, 1 g of the product covers a surface area of 400 m^2. Many studies have shown that chitin nanofibrils can activate the proliferation of keratinocytes as well as fibroblasts, regulating not only collagen synthesis but also cytokine secretion and macrophage activity [12].

16.6.2 Nanoparticles as UV Protective Filters in Sun Screens

The microparticles of titanium dioxide and zinc oxide have been used in cosmetic products as UV filters. The substances were used as traditional white pigments in the micrometer sector. The resulting products however were comparatively thick, sticky pastes which were difficult to administer and were not widely adopted by consumers as they left a whitish film on the skin. By using nanoparticle-sized titanium dioxide and zinc oxide the substances are transparent to the human eye, can be applied more easily on the skin and leave a better skin feeling. In addition, protection against UV radiation has been radically improved as a result of the smaller sized particles. The primary size of the nanoparticles used as UV filters is approximately 40 nm. According to the German Cosmetic, Toiletry, Perfumery and Detergent Association (IKW) titanium dioxide and zinc oxide are now only to be found in nanoparticle form in sunscreens, and products carry a notice advising of the titanium dioxide and zinc oxide content. Currently only titanium oxide is listed as an authorized sunscreen filter on the EU Directive on cosmetics list of permitted UV filters [13].

References

1. Bettinger CJ, Borenstein JT, Khademhosseini A (2012) Introduction to the special section on nanostructured materials for tissue regeneration. IEEE T Nanobiosci 11:1–2
2. Meyyappan M (2006) Nanotechnology in aerospace applications. In: Nanotechnology Aerospace Applications 2007:7-1–7-2
3. Silva VA, de Castro LF, Candido GM et al (2013) Nanostructured composites based on carbon nanotubes and epoxy resin for use as radar absorbing materials. Mater Res 16:1299–1308
4. Petrov VM, Gagulin VV (2001) Microwave absorbing materials. J Inorg Mater 37:93–98
5. Wilcoxon JP, Abrams BL (2006) Synthesis, structure and properties of metal nanoclusters. J Chem Soc Rev 35:1162–1194
6. Diao Z, Kraus M, Brunner R et al (2016) Nanostructured stealth surfaces for visible and near-infrared light. Nano Lett 16:6610–6616
7. Younesi SR, Urbonaite S, Bjorefors F et al (2011) Influence of the cathode porosity on the discharge performance of the lithium-oxygen battery. J Power Sources 196:9835–9838
8. Zhang GQ, Zheng JP, Liang R et al (2010) Lithium-air batteries using SWNT/CNF buckypapers as air electrodes. J Electrochem Soc 157:A953–A956
9. Rea I, Terracciano M, Politi J et al (2015) Natural and synthetic nanostructured materials for biomedical applications. In: AEIT international annual conference (AEIT), pp 1–6
10. Baranwal A, Srivastava A, Kumar P et al (2018) Prospects of nanostructure materials and their composites as antimicrobial agents. Front Microbiol 9:422
11. Percot A, Viton C, Domard A (2003) Optimization of chitin extraction from shrimp shell. Biomacromol 4:8–18
12. Morganti P (2010) Use and potential of nanotechnology in cosmetic dermatology. Clin Cosmet Investig Dermatol 3:5–13
13. Greßler S, Gazso A, Simko M (2010) Nanotechnology in cosmetics. Nano Trust dossiers

Chapter 17
Nanostructured Materials Life Time and Toxicity Analysis

Abstract Now-a-days, the nanostructured materials increased its fabrication skill and application in important sectors like energy, environment, and medicine. However, there are some concerns related to its toxicity and life time. This chapter presents some interesting techniques and analysis to measure life time and toxicity of nanoscale materials.

The increasing use of nanomaterials in consumer and industrial products has stimulated global concern regarding their fate in biological systems, resulting in a demand for parallel risk assessment. A number of studies on the effects of nanoparticles in in vitro and in vivo systems have been published. However, there is still a need for further studies that conclusively establish their safety/toxicity, due to the many experimental challenges and issues encountered when assessing the toxicity of nanomaterials. Most of the methods used for toxicity assessment were designed and standardized with chemical toxicology in mind. However, nanoparticles display several unique physicochemical properties that can interfere with or pose challenges to classical toxicity assays. Recently, some new methods and modified versions of pre-existing methods have been developed for assessing the toxicity of nanomaterials [1].

The main focus of current nanomaterial toxicity research is engineered nanoparticles, such as metals, metal oxides, single-walled and multiwalled carbon nanotubes, C-60, polymeric nanoparticles used as drug carriers, and quantum dots. The increase in relative surface area that occurs as particle size decreases down to the nanoscale gives rise to novel and enhanced material properties, but it also renders them more biologically reactive. Reducing particles to nanosize can also give them access to distal regions of biological systems that are normally inaccessible to larger particles. The release of nanoparticles into the environment can occur through many processes, such as spilling and washing consumer products incorporating nanoparticles; during synthesis and production; as an accidental release during transport or use; from industries that exploit nanotechnology, for example wastewater treatment and drug delivery [2].

T. D. Thangadurai et al., *Nanostructured Materials*, Engineering Materials,
https://doi.org/10.1007/978-3-030-26145-0_17

17.1 Impact of Nanomaterials to Human Health and Ecosystems

Considering the assessment of nanomaterials life cycle, one of the main concerns is if they exert any potentially negative impact on human health and ecosystems. Potential impacts of nanoparticles may be immensely underestimated if genotoxicity is not considered. For this reason, toxicity, i.e. genotoxicity and ecotoxicity have been considered. It is shown that the exposure to nanoparticles in plants induces reduced germination or growth, membrane damage, impaired photosynthesis, slowed or reduced reproductive development and mortality. The exposure to nanoparticles in animals induces cytotoxicity by necrosis or apoptosis, tissue or organ-level damage, growth inhibition, impaired reproduction and/or development and mortality [3].

For humans, molecular changes upon exposure can affect either somatic or germ cells and subsequent adverse birth outcomes and genetic diseases and carcinogenesis. The mitotic spindle aberrations are anticipated in exposed workers to SWCNT. The observed disruption is common in many solid tumors including the lung cancer. Cytotoxic and genotoxic properties of SWCNT have been verified in cells of the human gastrointestinal tract. Also, oxidative stress responses in the lungs of mice can be triggered by inhalation of pure CNT. Some recent studies found that CdSe-core QDs were indeed toxic under certain conditions and this toxicity can be modeled by processing parameters during synthesis, UV light exposure and surface coating and that it correlates with the liberation of Cadmium ions. In some studies it is shown that the penetration of quantum dots through abrasions in the skin is possible, and also that quantum dots can be transferred through a small food chain. The minute concentrations of quantum dots could be sufficient to cause long lasting, even trans generational, effects in exposed cells. Even though quantum dots can partially degrade in the environment or in biological systems over time, they can eventually cause small, but cumulative undesirable effects. Regarding nano-TiO_2, there are a lot of uncertainties and discrepancies in the achieved results. The investigation of potential genotoxicity of TiO_2 nano-particles exposure showed contradictory results. After 5 days inhalation of mice, no genotoxic effects were observed while, on the other hand, it is also showed that TiO_2 nanoparticles can induce genotoxic affects both in vivo and in vitro tests. In vivo tests showed that TiO_2 nanoparticles can enter directly into the brain through the olfactory bulb and can be deposited in the hippocampus region, damaging rat and human glial cells. The investigation of the potential ecotoxicity impacts on algae, daphnia and fish as a result of direct release of Ag and TiO_2 nano-particles (mainly <200 nm in nominal diameter size) from various nano-materials products to the freshwater has been analyzed. It was shown that nanomaterials, constituted from TiO_2 had lower ecotoxic impact than those made from Ag and there was a linear regression between Ag nanomaterials content in the considered products and the potential ecotoxicity impacts to the freshwater species, according to the release of total Ag during use (mainly washing). In general, if and how genotoxicity of nanoparticles can be induced in higher trophic level organisms through food chain remain undiscovered [4].

The current understanding of Engineered Nanoparticles (ENP) fate and transport in environmental and biological systems is poor and the current literature relies exclusively on nominal exposure. The results of potential toxicity, i.e. genotoxicity of nanoparticles in the living organisms, like for nano-sized TiO_2, are taken from certain number of studies that deal with the estimation based on the results gathered from in vitro or in vivo studies where no origin of these particles is taken into account, i.e. the pathways of these particles from their release till their ingestions by living beings remain unknown. In other words, the indications of genotoxicity considered in these studies are not strictly associated with the emissions of nano-TiO_2 from the structures supporting renewable energy provision and storage, such as nanostructured titania catalyst system and other similar structures, but represents undefined and more general situation. Due to the lack of exact data about behavior and impact on human health of these particles, it can only be predicted that they are causing the same effects on human and animal genome as TiO_2 nano particles studied in many articles [5].

17.2 Nanomaterial Toxicity

Humans are exposed to nanoparticles as they are produced by natural processes. Production, use, disposal, and waste treatment of products containing nanoproducts are the prime reasons for the environmental release of nanoparticulates in the original or modified forms. Foreign substances are generally blocked by human skin, whereas organs susceptible to foreign substances include the lungs and gastrointestinal tract. Nanoparticles (NPs) are comparable to viruses in size. For instance, the diameter of the human immunodeficiency virus (HIV) particle is on the order of 100 nm. NPs that are inhaled can effortlessly reach the bloodstream and other sites in the human body including the liver, heart or blood cells. It is significant to mention that the toxicity of NPs depends on their origin. Many of them seem to be nontoxic and others have positive health effects. The small size of NPs facilitates translocation of active chemical species from organismal barriers such as the skin, lung, body tissues and organs. Thus, irreversible oxidative stress, organelle damage, asthma, and cancer can be caused by NPs depending on their composition. The general acute toxic effects caused by exposure to NPs and nanostructured materials include reactive oxygen species generation, protein denaturation, mitochondrial disconcertion and perturbation of phagocytic functions. Uptake by the reticuloendothelial system, nucleus, neuronal tissue and the generation of neoantigens that causes possible organ enlargement and dysfunction are common chronic toxic effects of NPs [6].

17.3 A Consideration of All Pertinent Sources of Nanomaterials

The production of engineered nanomaterials (ENMs) creates an immediate concern; due to its risk assessment for nanomaterials must take into account with the relative magnitude of ENMs as sources compared with other sources of materials that may be identical or similar to ENMs. Nature produces a plethora of nano-scale particles in processes ranging from forest fires to bacterial metabolism. Human activities may also produce nano-scale particles by precipitation in waste streams, internal combustion engines, and other "incidental" sources. In some cases the materials produced are identical to ENMs as in the case of fullerenes produced in engineered, natural, or incidental combustion processes [7].

Incidental carbon nanotubes (CNTs) and other fullerene-related nanocrystals have been reported to originate from propane stoves, wood fires, burning tires and other sources and fullerene C_{60} has been found in geologic deposits, candle soot, and meteorites. TiO_2 nanoparticles, similar to ENMs, have been found downstream of hazard waste sites. An assessment of exposure to nanomaterials must also address possible releases associated with various stages of fabrication, transport, processing and disposal; activities that make up what is referred to as the *value chain* of nanomaterial production and use. The nanomaterial value chain involves the production of basic building blocks of nanomaterials and their incorporation in subsequent stages into products of increasing complexity [8]. For example, engineered nanomaterials such as titanium dioxide might be modified with a tailored surface chemistry to yield suspensions that are then used to create various products ranging from thin films for self-cleaning windows to catalysts suspensions in water treatment. At each stage in the value chain, there exists the possibility of nanomaterial release and subsequent exposure to humans or ecosystems through the production, transport, use and disposal of nanomaterials and nanomaterial-containing products. Important factors to be identified in evaluating potential nanomaterial exposure at each stage in the value chain are the format that nanomaterials will be present in as commercial products, the potential for these materials to be released to the environment, and the transformations that those materials may undergo that may affect their subsequent potential for exposure. Indeed, due to modifications along the value chain or environmental transformations, the potential contact between humans and ecosystems outside of the work place will most likely involved nanomaterials there bear little resemblance to the initial material [9] (Fig. 17.1).

An additional set of hazards to be examined is the potential for collateral damage i.e. environmental impacts that arise from the production and use of nanomaterials rather than the nanomaterials themselves. In particular, the environmental impacts associated with upstream energy usage are likely to contribute significantly to the environmental footprint of nanomaterials production given the high energy inputs currently employed to create order at the nanoscale. These issues may greatly outweigh direct health or environmental impacts associated with an emerging nanomaterials industry. The first ever published work on nanomaterial risk assessment dealt

Fig. 17.1 Evaluation of nanomaterial risks to ecosystems and organisms posed by multiple sources

directly with issues of collateral damage. One of the important findings reported in this work was that methods for manufacturing nanomaterials tend to become greener with time; substituting, for example, less toxic solvents or implementing more energy-efficient procedures for fabricating nanomaterials. Subsequent work by others looking at carbon nanotube production has shown that nanomaterial production may involve the production of non-nano wastes that pose significant hazards [10].

17.4 Nanoparticle Toxicity

When bulk materials are made into nanoparticles, the surface to volume ratio of the material increases. When this process reaches the nanoscale, the proportion of surface atoms or molecules in the material increases exponentially and the surface chemistry also changes, with the material tending to become more chemically reactive. This is the basis for the production of heterogeneous catalysts in the chemical industry. Platinum, for example, in the bulk state is particularly chemically unreactive. In the form of ultrafine particles, however, it can facilitate a number of chemical reactions [11]. The term nanotoxicology, first used by Donaldson [11] indicates the fact that such differences require an ad hoc form of toxicology. The essential nature of the nanotoxicological problem was captured by Myllynen [12] "There are basically two questions to address: (1) what is the fate of nanoparticles? (i.e. where do they get to) and (2) if they get to a particular location, does it matter?".

17.4.1 Mechanisms of Toxicity

A number of mechanisms by which nanoparticles can induce cell damage have been reported, including, for example:

- oxidative damage through catalysis
- lipid peroxidation
- surfactant properties
- protein misfolding
- direct physical damage
- enzyme poisoning [12].

17.5 Interference of Nanoparticles with in Vitro Toxicity Assays

In vitro experimentation has always been the first choice for toxicologists, since it is time- and cost-effective. Although it cannot replace animal experimentation completely, but it does help to ensure that they are only used when absolutely necessary, and it sometimes provides mechanistic information on the toxicity of nanoparticles after in vivo studies. The risk assessment of different aspects of nanotechnology is still in its early stages. Therefore, most of the studies pertaining to nanoparticle toxicology that have been carried out so far have been preliminary and confined to the classical in vitro toxicity test methods established for drugs and chemicals [13]. However, the methods that are used in traditional toxicology cannot be applied present nanoparticle toxicology, as nanoparticles display several unique physicochemical properties. Due to these properties, nanoparticles interfere with normal test systems, and this interference has been well documented. Examples of such properties include: high surface area, leading to increased adsorption capacity; different optical properties that interfere with fluorescence or visible light absorption detection systems; increased catalytic activity due to enhanced surface energy; and magnetic properties that make them redox active and thus interfere with methods based on redox reactions [14] (Fig. 17.2).

17.6 Nanotoxicology

The distinctive and various physicochemical properties of nanoscale materials recommend that their toxicological properties may differ from those of the corresponding bulk materials. The potential occupational and public exposure through inhalation, oral ingestion, dermal absorption or by injection of manufactured nanoparticles (mNPs) with particles size $\leq$ 100 nm probably will increase in the near future, due

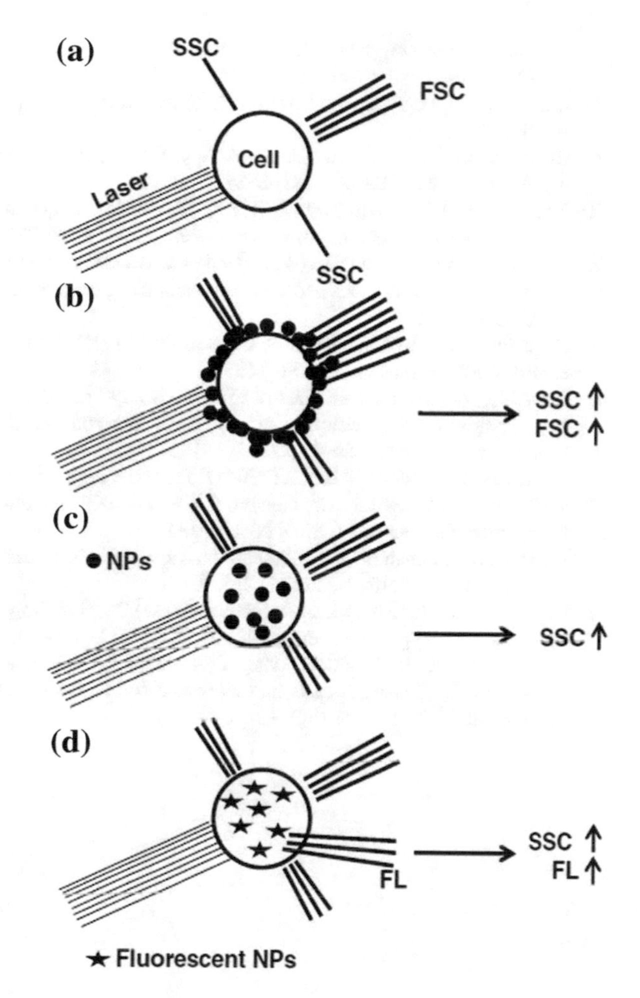

Fig. 17.2 a–d Analyzing nanoparticle (NP) uptake in cells by flow cytometry: **a** light scattering by a cell that is not associated with any nanoparticle; **b** nanoparticles adhere to the cell surface, leading to an increase in forward scatter (FSC) and side scatter (SSC); **c** nanoparticle internalization by the cell, leading to an increase in SSC alone; **d** fluorescent nanoparticle internalization by the cell, leading to an increase in SSC and fluorescence intensity (FL)

to the ability of nanomaterials to improve the quality and performance of many consumer products as well as the development of therapeutic strategies and tests. However, there is still a lack of information about the impact of mNPs on environment and on human health, as well as of reliable data on risk assessment [15].

References

1. Dhawan A, Sharma V (2010) Toxicity assessment of nanomaterials: methods and challenges. Anal Bioanal Chem 398:589–605
2. Kroll A, Pillukat MH, Hahn D et al (2009) Current in vitro methods in nanoparticle risk assessment: limitations and challenges. Eur J Pharm Biopharm 72:370–377
3. Wang H, Wu F, Meng W et al (2013) Engineered nanoparticles may induce genotoxicity. Environ Sci Technol 47:13212–13214

4. Winnik FM, Maysinger D (2013) Quantum dot cytotoxicity and ways to reduce it. Acc Chem Res 46:672–680
5. Chen T, Yan J, Li Y (2014) Genotoxicity of titanium dioxide nanoparticles. J Food Drug Anal 22:95–104
6. Gnach A, Lipinski T, Bednarkiewicz A (2015) Upconverting nanoparticles: assessing the toxicity. Chem Soc Rev 44:1561–1584
7. Robichaud CO, Uyar AE, Darby MR (2009) Estimates of upper bounds and trends in Nano-TiO_2 production as a basis for exposure assessment. Environ Sci Technol 43:4227–4233
8. Becker L, Poreda RJ, Nuth JA (2006) Fullerenes in meteorites and the nature of planetary atmospheres. Natural fullerenes and related structures of elemental carbon. Springer, Netherlands, pp 95–121
9. Hochella MF, Lower SK, Maurice PA et al (2008) Nanominerals, mineral nanoparticles, and earth systems. Science 319:1631–1635
10. Plata DL, Gschwend PM, Reddy CM (2008) Industrially synthesized single-walled carbon nanotubes: compositional data for users, environmental risk assessments, and source apportionment. Nanotechnol 19:185706–185719
11. Donaldson K, Stone V, Tran CL (2004) Nanotoxicology. Occup Environ Med 61:727–728
12. Myllynen PK, Loughran MJ, Howard CV et al (2008) Kinetics of gold nanoparticles in the human placenta. Reprod Toxicol 26:130–137
13. Kahru A, Savolainen K (2010) Potential hazard of nanoparticles: from properties to biological and environmental effects. Toxicol 269:89–91
14. Doak SH, Griffiths SM, Manshian B et al (2009) Confounding experimental considerations in nanogenotoxicology. Mutagenesis 24:285–293
15. Bergamaschi E, Bussolati O, Magrini A et al (2006) Nanomaterials and lung toxicity: interactions with airways cells and relevance for occupational health risk assessment. Int J Immunopathol Pharmacol 19:3–10

Chapter 18
Nanomaterials Research and Development

Abstract The specific impact caused by nanoparticles can be assessed by life cycle assessment analysis. Performing life cycle analysis and potential risk management of nanoscale materials are increasing to perform better products and further technology developments. This chapter describes methods for research developed and applied life cycle assessment (LCA) models to evaluate ENMs and nano-enabled product.

18.1 LCA—Life Cycle Assessment of Nanomaterials

Life cycle assessment (LCA) is based on the conception of life cycle, it is a systematic method for assessing the potential environmental impacts of products, services and processes across their entire life cycles. The LCA methodology, as defined in ISO 14040, comprises four iterative steps: (i) goal and scope definition, (ii) inventory analysis, (iii) impact assessment, and (iv) interpretation. LCA is accepted as a powerful tool to identify potential life cycle impacts of nanotechnologies [1].

18.2 The Role of Life Cycle Assessment in the Field of Nanotechnology

Nanotechnology and the production of nanomaterials and products containing nanomaterials are known as nanoproducts; however, numerous uncertainties exist regarding their possible impact on the environment and human health. Therefore, holistic and comprehensive assessment tools such as Life Cycle Assessment (LCA) are essential to analyze, evaluate, understand and manage the environmental and health effects of nanotechnology. The knowledge of the exposure routes as well as of the potential environmental impacts of nanoparticles is limited. In addition, potential resource and environmental advantages of nanomaterials and products using nanomaterials over conventional products have not been investigated. Therefore, a clear need exists to establish a full understanding of the environmental benefits and drawbacks of nanotechnology and nanomaterials compared with those of conventional technologies

T. D. Thangadurai et al., *Nanostructured Materials*, Engineering Materials,
https://doi.org/10.1007/978-3-030-26145-0_18

and products over their complete life cycles. LCA is the essential tool to achieve this [2]. The goal and scope definition of an LCA provides a description of the product system in terms of the system boundaries and a functional unit, i.e., the reference unit defining the function of the product system.

The Life Cycle Inventory (LCI) aims at data collection and calculation procedures in order to quantify from cradle-to-grave the relevant inputs (e.g., material inputs) and outputs (e.g., emissions to air) of the product system. The purpose of the Life Cycle Impact Assessment (LCIA) is to aggregate the results from the inventory analysis and to evaluate the significance of the product's potential environmental impacts. This process involves connecting inventory data with specific environmental impact categories and the respective category indicators, such as global warming potential as an indicator for climate change. The interpretation considers the findings from both LCI and LCIA and should provide conclusions and recommendations [3] (Fig. 18.1).

Note that LCA is dissimilar from other techniques, such as environmental performance evaluation and risk assessment (RA), as it is a relative and essentially comparative approach based on a functional unit, with all inputs and outputs accounted for in the LCI, and consequently in the LCIA profile, being associated with the functional unit. LCA is adequate to answer many, but certainly not all, questions on environmental and human-health impacts of nanotechnology in comparing different products with the same function. However, for assessing social and economic benefits and/or problems, as would be required in a full sustainability assessment, a broader assessment framework is required [4].

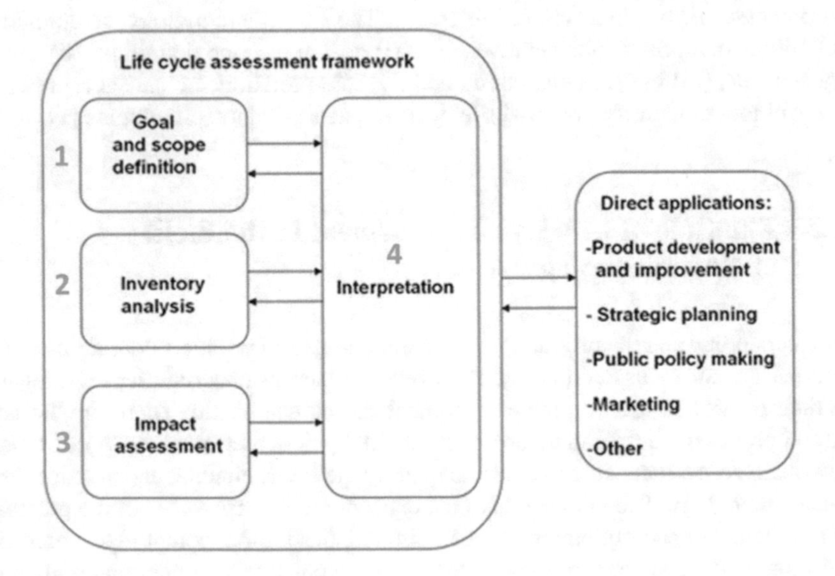

Fig. 18.1 Stages of life cycle assessment

18.3 LCA Procedure

It consists of four steps: scoping and definition of system boundaries, inventory of material and energy requirements, environmental impact analysis, and improvement analysis an iterative procedure that takes place continuously as the LCA is carried out. LCAs are undertaken for four principal reasons:

(i) to develop information for comparative purposes when choices among designs, processes, or materials must be made,
(ii) to quantitatively establish the nature of tradeoffs associated with various choices,
(iii) to examine a single system in order to ascertain those components, parts, or processes which are the most material or energy intensive and for which investments of resources or research might be expected to yield the greatest improvements, and
(iv) to facilitate the communication of risks and benefits to consumers and stakeholders [5].

LCA impact analysis proceeds by assembling a material inventory of chemical quantities released to the environment throughout the supply chain. The result is typically a list of impact categories, such as global warming or human toxicity potential, that are relevant to the study. Characterization proceeds by converting each chemical quantity into an equivalent amount of substance that is represent of the impact category, such as kg of carbon dioxide (for global warming), or benzene equivalents (for cancer). In the case of nanomaterials, the use of mass-based equivalencies may be inappropriate, i.e. nanomaterials may be better characterized in terms of their principal functional property, which necessitates an understanding of the form of these materials in both the product and the environment, and their rate of release [6].

18.4 Life-Cycle Assessment of Engineered Nanomaterials

The progressive diffusion of ENMs in many fields, including nanoremediation, and the global consensus that their release into the environment will increase, has led not only to the urge of a sound evaluation of their toxicity effects on human health and on the environment, but also to the need for the evaluation of their environmental sustainability [7]. Life-cycle assessment (LCA) is a well-established tool, nowadays largely used to evaluate the potential environmental impacts of a product system over its whole life cycle, from the extraction and acquisition of raw materials, to the core production process, use and end of life treatment, either recovery or final disposal. The method consists in the compilation and evaluation of the inputs, outputs, and the potential environmental impacts of a product system throughout its entire life cycle, from cradle to grave. LCA is considered a holistic method since it provides the assessment of the potential environmental impacts on several environmental categories, mainly on global and regional scale, such as global warming potential, ozone depletion potential, acidification potential, resource depletion etc. Moreover, LCA permits to define the environmental hotspots of a product system, to analyze alternative solutions that provide performance improvement and to make comparison of different scenarios, therefore proving to be a powerful tool for supporting eco-design and decision-making. LCA is deemed to be the suitable tool to assess the environmental impacts of emerging technologies such as nanotechnologies and nano-enabled products, also in comparison to conventional technologies. Although LCA is strongly recommended as tool to assess the sustainability of ENMs throughout their life cycle, the scientific community currently agrees on the several information gaps, which hamper the proper application of LCA in the field. These gaps regard mainly two broad issues, namely the difficulty of including the whole life cycle of ENMs and to fully assess their impacts on human toxicity and ecotoxicity.

The first issue stems from the lack, in the life-cycle inventory, of specific features and properties of the new nanomaterials that differentiate them from the corresponding bulk material and of the quantification of their release into the environment across their life-cycle [8]. The second issue regards the application of the impact assessment methods, which currently do not allow consideration of the nano specific impacts on human health and ecotoxicity. In fact, on one hand the fate of ENMs in the environment is poorly modeled and quantitatively assessed and on the other hand, current impact assessment tools (e.g., USEtoxTM) still lack to consider nano specific properties and thus are not able to provide suitable characterization factors to include them in the assessment of impacts on health and environment [9].

18.5 LCA of an Emerging Technology

LCA to emerging technologies due to the lack of detailed knowledge regarding the inputs and output of the system. Nevertheless, there is a general trend to apply LCA to emerging technologies (e.g., solar, wind, bio-fuels). In the case of CNTs, note that the application of LCA is a challenging task because many of the technologies studied are still emerging; introducing a great degree of uncertainty and complexity into LCA. Obtaining accurate data for emerging technologies can be a challenge because data based on conceptual designs, and assumption about the scaling up of laboratory or pilot scale process may not accurately reflect industrial scale operations. Additionally, early prototypes may undergo several changes during product development and testing that can alter how a product is manufactured and used. Furthermore, one of the greatest challenges when assessing nanoproducts is the variable nature of manufacturing processes and how subtle differences in the resulting nano components can affect the associated nanoproduct [10] (Fig. 18.2).

Increasing research efforts have been made development in the methods to synthesize nanomaterials without using raw materials containing scarce natural resources and hazardous substances, or the production methods that require low energy and low material consumption. The concept of green nanotechnology applies not only to

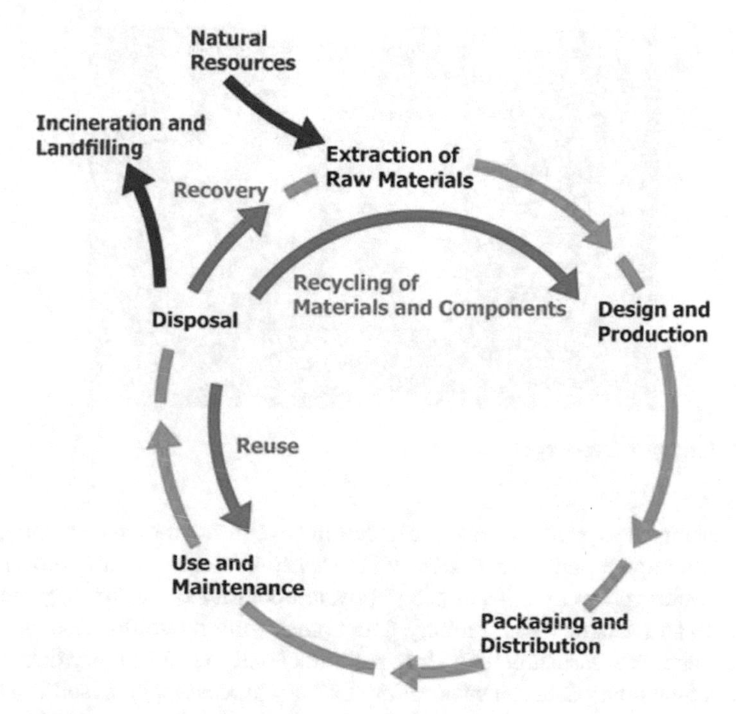

Fig. 18.2 Life cycle of a product system. Reprinted with permission © UNEP (2007)

the manufacturing stage but also to the usage stage, in particular, in environmental applications. Caution is required so as not to consider the technology as green, only based on the fact that applications are in the environmental and renewable-energy sectors. The potential environmental burden of the applications during manufacturing and disposal stages should be taken into account. Life Cycle Thinking (LCT) helps in guiding the nanotechnology development towards true eco-friendliness, through the assessment of the overall risk-benefit balance during the production, usage and disposal stages of nano-products. Nanotechnology is by nature a multidisciplinary research field, but the LCT of nanotechnology calls for strong collaborations from the wider research community beyond physics, chemistry, medicine and biology [11].

18.6 Life Cycle Inventory (LCI)

The inventory analysis comprises the data collection and the calculation procedures to quantify the inputs energy, raw and ancillary materials, water, etc., and outputs emissions to soil, emissions to air, water emissions, and waste water and products/sub products through the system boundaries. Each life stage is analyzed to determine the relevant inputs and outputs of the system, performing a flow balance. To make the

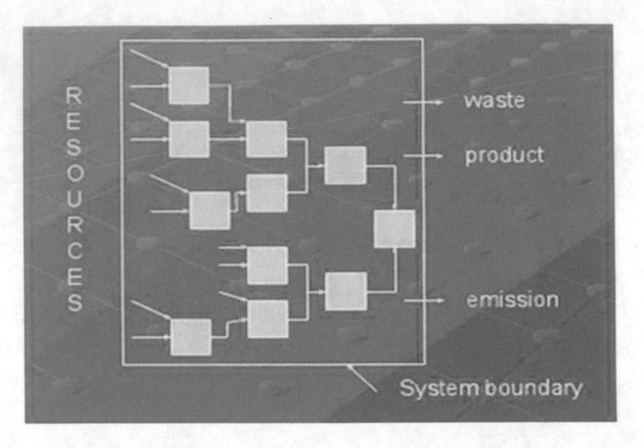

Fig. 18.3 Life cycle inventory (LCI)

process easier, the system is usually divided in several interconnected subsystems. Quality data requirements are necessary in order to guarantee their appropriateness and representativeness in terms of geography, temporality and technology, information source and accurateness. Primary data comes from modeling/monitoring processes through real measurements. It is recommended to use primary data for core processes. Secondary data can be used for auxiliary processes; main sources for secondary data are literature and databases (such as the ELCD, European reference Life Cycle Database, or commercial databases like Ecoinvent or GABI) [12] (Fig. 18.3).

18.7 Life Cycle Impact Assessment (LCIA)

LCIA is the estimation of indicators of the environmental pressures associated with the environmental interventions attributable to the life-cycle of a product. In this step, the LCA the inventory flows are converted into the associated potential environmental impacts. LCIA stage has four steps:

1. Classification (Mandatory): Assignation of the different material/energy inputs and outputs inventoried to the relevant impact categories.
2. Characterization (Mandatory): Calculation of the magnitude of the contribution of each classified input/output to their respective impact categories and aggregation of the contributions within each category.
3. Normalization (Optional): impact assessment results are multiplied by normalization factors (e.g., European reference values) in order to calculate and compare the magnitude of their contributions to the impact categories in a dimensional way.
4. Weighting (Optional): in order to support the interpretation of results, normalized results are multiplied by a set of weighting factors which reflect the perceived relative importance of the impact categories considered [13] (Fig. 18.4).

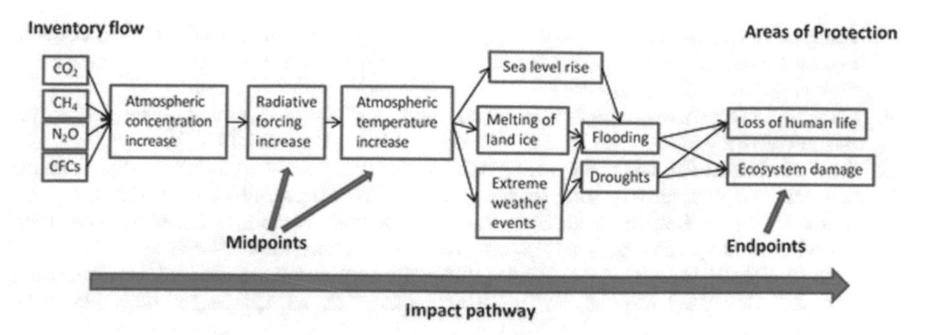

Fig. 18.4 Life cycle impact assessment mechanisms and steps [16]

18.8 LCA of Nanomaterials

Life cycle assessment of nano-based technologies are important to detect all potential environmental burdens, whether positive or negative, of nano-products and to prevent and treat all potential environmental risks of further technological developments. So far, a certain number of LCA studies dealing with the use of NM in sector of renewable energy have been conducted. The most studied nanomaterials in the aspect of LCA methodology application are carbon nanotubes, i.e. single-walled carbon nanotubes (SWNT), which are applied in Li-ion batteries, nano photovoltaic, such as quantum dots and dye-sensitized solar cells, nano-TiO_2 which are used in solar PV cells and for hydrogen generation. Of all environmental impacts occurring from nanomaterials life-cycle toxicity is the one that poses most concern [14]. LCA approach for nanotechnology and nano-products can provide useful information about the main environmental impacts and benefits of this emerging technology. Prospective LCA approaches are needed and experimental data on characteristics and toxicity of nanoparticles coming from research projects should be included in LCA methodologies. Adapted exposure and fate modeling are needed in order to have complete results on the environmental performance of nano-products during all life cycle stages. LCA information should be used together with other methodologies such as Risk Assessment (RA) to obtain a deep comprehension on the interactions of nanomaterials and the environment and the potential damage on environment and human health in all life cycle stages and exposure levels [15].

References

1. Salieri B, Turner DA, Nowack B et al (2018) Life cycle assessment of manufactured nanomaterials: where are we? NanoImpact 10:108–120
2. Grieger KD, Laurent A, Miseljic M et al (2012) Analysis of current research addressing complementary use of life-cycle assessment and risk assessment for engineered nanomaterials have lessons been learned from previous experience with chemicals? J Nanoparticle Res 14:958–980

3. Tsang MP, Sonnemann GW, Bassani DM (2016) A comparative human health, ecotoxicity, and product environmental assessment on the production of organic and silicon solar cells. Prog Photovolt: Res Appl 24:645–655
4. Villares M, Isildar A, Mendoza Beltran A et al (2016) Applying an ex-ante life cycle perspective to metal recovery from e-waste using bioleaching. J Clean Prod 129:315–328
5. Theis TL, Bakshi BR, Durham D et al (2011) A life cycle framework for the investigation of environmentally benign nanoparticles and products. Phys Status Solidi RRL 5:312–317
6. Auffan M, Rose J, Bottero J et al (2009) Towards a definition of inorganic nanoparticles from an environmental, health and safety perspective. Nat Nanotechnol 4:634–641
7. Corsi I, Fiorati A, Grassi G et al (2018) Environmentally sustainable and ecosafe polysaccharide-based materials for water nano-treatment: an eco-design study. Materials 11:1228–1250
8. Gavankar S, Suh S, Keller AF (2012) Life cycle assessment at nanoscale: review and recommendations. Int J Life Cycle Assess 17:295–303
9. Bauer C, Buchgeister J, Hischier R (2008) Towards a framework for life cycle thinking in the assessment of nanotechnology. J Clean Prod 16:910–926
10. Lazarevic D, Finnveden G (2013) Life cycle aspects of nanomaterials. US AB, Stockholm
11. Tsuzuki T (2014) Life cycle thinking and green nanotechnology. Austin J Nanomed Nanotechnol 2:1
12. Seager TP, Linkov I (2009) Uncertainty in life cycle assessment of nanomaterials. Nanomaterials: risks and benefits. Springer, Netherlands, pp 423–436
13. Hischier R, Walser T (2012) Life cycle assessment of engineered nanomaterials: state of the art and strategies to overcome existing gaps. Sci Total Environ 425:271–282
14. Peric M, Hut I, Pelemis S et al (2015) Possible approaches to LCA methodology for nanomaterials in sustainable energy production. Contemp Mater Renew Energy Sources 2:161–169
15. Hidalgo C, Gonzalez-Galvez D, Janer G et al (2013) Life cycle assessment (LCA) of nanomaterials: a comprehensive approach. In: 6th international conference on life cycle management, Gothenburg
16. Hauschild MZ, Huijbregts MAJ (2015) Introducing life cycle impact assessment. In: Hauschild M, Huijbregts M (eds) Life cycle impact assessment. LCA compendium—the complete world of life cycle assessment. Springer, Dordrecht

Printed in the United States
by Baker & Taylor Publisher Services